T0233695

SpringerBriefs in Computer Science

More information about this series at http://www.springer.com/series/10028

Jaka Sodnik • Sašo Tomažič

Spatial Auditory
Human-Computer Interfaces

Springer

Jaka Sodnik
Faculty of Electrical Engineering
University of Ljubljana
Ljubljana, Slovenia

Sašo Tomažič
Faculty of Electrical Engineering
University of Ljubljana
Ljubljana, Slovenia

ISSN 2191-5768   ISSN 2191-5776 (electronic)
SpringerBriefs in Computer Science
ISBN 978-3-319-22110-6   ISBN 978-3-319-22111-3 (eBook)
DOI 10.1007/978-3-319-22111-3

Library of Congress Control Number: 2015945657

Springer Cham Heidelberg New York Dordrecht London

Printed on acid-free paper

Springer International Publishing AG Switzerland is part of Springer Science+Business Media (www.springer.com)

Abstract

Human–computer interfaces enable the exchange of information between human users and various types of machines, computers and other electronic devices. The interaction with modern devices is most commonly performed through visual, auditory or tactile user interfaces. In the case of an auditory interface, the information is presented with different types of sounds. Auditory interfaces can complement visual interfaces or in some cases represent an independent input or output interface to a selected device.

This survey focuses on a special group of auditory interfaces using spatial sound for the representation of information. The addition of information on the location of a selected sound source or a group of sources shows many advantages over a mere single-channel audio. This survey explains the most important limitations of the human hearing system and the perception of spatial sound. It also includes some technical background and basic processing and programming techniques for the creation and reproduction of spatial sounds with different audio equipment.

Spatial auditory interfaces have evolved significantly in the last couple of years and can be found in a variety of environments where visual communication is obstructed or completely blocked by other activities, such as walking, driving, flying, operating multimodal virtual displays, etc. Separate chapters of this survey are dedicated to the most important areas of spatial auditory displays: mobile devices and computers, virtual environments, aircrafts and vehicles, visually impaired and blind computers users and brain–computer interfaces.

Keywords Human–computer interaction • User interface • Spatial audio • Localization • HRTF

Contents

Chapter 1
Introduction

Research and development work in the field of electrical engineering and other technical sciences has always aimed to develop more powerful devices with ever-increasing processor power and responsiveness (Sodnik et al. 2014). A reduction in the size of modern devices and the extension of their use to new fields and environments has been a significant design trend in recent years. Mobile devices enabling various types of communication and access to important sources of information in mobile environments constitute an important segment of modern electronics. Mobile devices can be general-purpose laptops, notebooks, tablets or smartphones, or can be highly-adapted purpose-built devices found in vehicles and other means of transport. In terms of the range of functionalities and the corresponding software, the majorities of advanced mobile devices are becoming very similar to traditional desktop computers, and thus allow for continuous upgrades and the addition of new applications.

The way an information and communication device is used is determined by the input–output components that make up its user interface. Human–Computer Interaction (HCI) or Human–Machine Interaction (HMI) is a scientific field that studies the various ways in which information and communications can be exchanged between people and devices of all sorts (Dix et al. 2004). It is an interdisciplinary field, and incorporates knowledge from electrical engineering, computing, mechanical engineering, psychology and cognitive science. The user interface is the sole point of contact between the device and its user, and must provide for effective control of the functioning of the device and the input of information and requests. It should also provide fast, transparent responses and feedback, based on which the user can evaluate the success of commands and orders, and thus decide on subsequent steps and actions. The user interface includes hardware and software, and consists of input and output elements. The input element allows the manipulation of the system by the user, while the output element reports feedback of the success of the manipulation and the effects on the actual system to the user.

© The Author(s) 2015
J. Sodnik, S. Tomažič, *Spatial Auditory Human-Computer Interfaces*,
SpringerBriefs in Computer Science, DOI 10.1007/978-3-319-22111-3_1

The communication channels available for this interaction are visual, audio and tactile. They are based on the three primary human senses through which we are able to take in the environment and the world around us. These are seeing, hearing and feeling. Individual senses differ greatly from one another, and therefore possess unique attributes and limitations that have to be taken into account when user interfaces are being designed. In particular, there is a clear difference in the available transfer rate, i.e., in the quantity of information that can be transmitted via an individual sense in a unit of time. Seeing has the highest capacity, allowing for the simultaneous intake of a large quantity of information, and has a relatively large scope and focus (Koch et al. 2006). Hearing and feeling have significantly less capacity. The hearing channel is multi-directional, and allows for the recognition of sounds from all directions, but does not allow for the effective recognition and distinction of a large number of simultaneous sources unless they are separated in space (Jacobson 1950).

Nowadays, the vast majority of HCI is based on visual communication, where information about a system is displayed on a screen. Older types of displays supported only the printout of alphanumeric characters or text in black-and-white or other monochrome technology (Myers 1998). Information flow was relatively poor and any more complex interaction was very time-consuming. More recent times brought the Graphical User Interface (GUI) in color, which could display more information and at the same time enhance the user experience with images and other multimedia elements. The GUI is still being developed, and today it can display various icons and objects in 3D. In any case, information flow is clearly increasing because of the higher resolutions and larger sizes of computer screens.

Sound or acoustic information appeared relatively early in the history of HMI and has seen hardly any change since (Sodnik 2007; Dicke 2012). The main motivation for the use of sound for interacting with machines is the high visual workload or even overload of the human visual system. Some pieces of information or warnings can simply be overlooked and ignored due to the high density of icons, images and texts in the visual display. Auditory signals can effectively and unobtrusively inform the user about some ongoing background processes or activities in the system without requiring his or her full attention and without complete distraction from another activity that has a higher priority (Sodnik et al. 2008a). They are mostly used:

- for events which require immediate user attention,
- for messages that are simple and short,
- as an alert message when the user's visual system is overloaded, and
- in cases where visual perception is impaired or impossible.

In some cases audio interfaces can represent a primary communication channel and completely substitute the visual interaction. An extreme example is the case of blind and partially-sighted users, who operate information and communication technologies solely by means of hearing and feeling. Examples of the interfaces available to them include screen readers for desktops and mobile devices, and special Braille keyboards, which convert screen content into tactile information. Users with unimpaired sight can often find themselves in very similar situations when their visual channel is overloaded and does not allow for efficient interaction with a mobile device. This happens when the user is mobile and needs the visual channel to perform a basic activity such as walking, running, cycling or driving. Other senses can also be overloaded in such instances, but undertaking the primary activity puts the greatest load on the visual channel. Secondary activities such as using a mobile phone or an inbuilt mobile device can present a distraction and endanger the user, and can significantly reduce the user's concentration. Auditory user interfaces are flexible and scalable, ranging from simple non-speech cues, to earcons, auditory icons, hearcons, natural or synthetic speech output, and audio representation of multivariate and multi-dimensional data compounds (Sodnik et al. 2008b). They do not interfere with visual information processing. In addition, audio signals can capture the user's attention even if they come from a certain distance or if their source is hidden from view. This makes auditory interfaces a good alternative in predominantly mobile domains.

On the other hand, the human auditory system has not evolved for accurate recognition of the environment or specific surrounding objects, but rather to alert humans to specific important changes or phenomena (Blauert 2001). Humans are terrible at perceiving and understanding a large number of simultaneous sounds, especially if they are speech sounds. Nevertheless, simultaneous perception of a large amount of audio information improves when individual sources are separated in space (Drullman and Bronkhorst 2000). The human brain is capable of determining the exact location of an individual sound source in space. The final localization accuracy depends on the spectral content of the signal (narrowband or broadband signals), relative location to the user (in front of, behind, to the side) and environment (a large room, an anechoic room, an open space, etc.). The localization mechanisms are based on the transfer function of the system from the origin of the source to the human eardrums. The accurate measurement of this transfer function is of vital importance for correct generation and reproduction of spatial sound. The possibility of reproduction of spatial sounds enables design and implementation of complex auditory interfaces for highly efficient interaction with various machines and devices. The information flow between a device and a user increases significantly according to spatial distribution of simultaneously-played sound sources.

In this book we review some basic principles and methods for the creation and reproduction of spatial sound and review some examples of its use in spatial auditory interfaces.

The first chapter offers an overview of basic properties of sound and its perception by human beings. First the phenomenon of the acoustic wave and its propagation properties is explained, followed by a description of the human auditory system and psychoacoustics. The main part of this chapter is the explanation of mechanisms of sound localization, their limitations and accuracy. Knowledge of localization is mandatory for the successful creation and reproduction of spatial sound through various speaker configurations. Publicly-available and widely-used 3D sound libraries are also listed and described briefly.

The second chapter describes general properties of auditory interfaces and there areas of use. Based on the sound-oriented classification they are divided into speech-based and non-speech interfaces. In addition to the description of their basic structures, properties and limitations the most commonly-used interface metaphors are also presented.

The third chapter is dedicated to spatial auditory interfaces proposed by individual researchers and renowned companies. They are grouped into several categories based on the area of use and dedicated interaction environment. The most important categories included in the review are:

- portable devices and music players,
- teleconferencing systems,
- augmented and virtual reality systems,
- aircrafts and vehicles,
- blind and visually impaired computer users and
- brain-computer interfaces.

The final selection of described interfaces is based on their scientific impact in the community, which is reflected by the large number of citations in various databases that discuss them.

References

Blauert J (2001) Spatial hearing, 3rd edn, The psychophysics of human sound localization. MIT, Cambridge, MA

Dicke C (2012) A holistic design concept for eyes-free mobile interfaces. Dissertation, University of Canterbury, Christchurch, New Zealand

Dix A, Finlay JE, Abowd GD, Beale R (2004) Human-computer interaction, 3rd edn. Pearson Education, Harlow

Drullman R, Bronkhorst AW (2000) Multichannel speech intelligibility and talker recognition using monoaural, binaural, and three-dimensional audtiory presentation. J Acoust Soc Am 107(4):2224–2235

Jacobson H (1950) The informational capacity of the human ear. Science 112(2901):143–144

Koch K, McLean J, Segev R, Freed MA, Berry MJ II, Balasubramanian V, Sterling P (2006) How much the eye tells the brain. Curr Biol 16(14):1428–1434

Myers BA (1998) A brief history of human computer interaction technology. ACM Interact 5(2):44–54

Sodnik J (2007) The use of spatial sound in human-machine interaction. Dissertation, University of Ljubljana, Slovenia

Sodnik J, Dicke C, Tomazic S (2008a) Auditory interfaces for mobile devices. In: Encyclopedia of wireless and mobile communications. Taylor & Francis, Boca Raton, FL, pp 1–9

Sodnik J, Dicke C, Tomažič S, Billinghurst M (2008b) A user study of auditory versus visual interfaces for use while driving. Int J Hum Comput Stud 66(5):318–332

Sodnik J, Kos A, Tomazic S (2014) 3D audio in human-computer interfaces. In: 3DTV-conference: the true vision-capture, transmission and display of 3D video (3DTV-CON), pp 1–4

Chapter 2
Spatial Sound

2.1 Acoustic Wave and Sound Propagation

Sound is a vibration which propagates as a mechanical wave through various types of compressible media (air, gas, liquids, or even solids). In most cases it is a longitudinal mechanical wave of pressure and displacement, although it can also have the form of a transverse wave in solids. The most important properties of sound are its amplitude, frequency (period), wavelength and speed of propagation (Kinsler et al. 1999). In a free field, sound is not subject to many influences and is therefore not reflected, absorbed, deflected or refracted in any way. The propagation of a sound wave is commonly described by the three-dimensional acoustic wave equation for sound pressure in the Cartesian coordinate system:

$$\frac{\partial^2 p}{\partial x^2} + \frac{\partial^2 p}{\partial y^2} + \frac{\partial^2 p}{\partial z^2} = \frac{1}{c^2}\frac{\partial^2 p}{\partial t^2} \tag{2.1}$$

where x, y and z denote the three coordinates in space, t denotes time, p denotes pressure and c denotes the speed of sound. The speed of sound in dry air at 20 °C at sea level is 343 meters per second (m/s) and it increases with the density of a medium or its temperature. The speed of sound in water is approx. 1500 m/s while the speed of sound in iron is almost 5200 m/s (Bamber 2004). The speed increases for approx. 0.6 m/s if the temperature of the air increases by 1 °C.

The movement of particles in a fluid medium results in energy transmission through this medium. The time-averaged rate of sound energy (W) transmission through a unit area is called sound intensity (I). It is defined as:

$$I = \frac{1}{T}\int_{0}^{T} pv\,dt \tag{2.2}$$

© The Author(s) 2015
J. Sodnik, S. Tomažič, *Spatial Auditory Human-Computer Interfaces*,
SpringerBriefs in Computer Science, DOI 10.1007/978-3-319-22111-3_2

where p denotes instantaneous pressure, v denotes the particle velocity and T denotes the period of the sound wave. The intensity is measured in Watts per square meter (W/m^2). For the harmonic plane wave, the intensity can also be expressed as:

$$I = \frac{p^2}{\rho_0 c} \tag{2.3}$$

where p again denotes pressure, ρ_0 denotes density of medium, and c denotes speed. As can be seen, the intensity is proportional to the square of the sound pressure.

In the case of a point source, the radiating sound is better presented in the spherical coordinate system (waves with spherical symmetry):

$$\frac{\partial^2 p}{\partial r^2} + \frac{2}{r}\frac{\partial p}{\partial r} = \frac{1}{c^2}\frac{\partial^2 p}{\partial t^2} \tag{2.4}$$

with p denoting pressure and r denoting the radial distance from the source. The source can be considered as a point due to its small physical dimensions compared with the distances where the radiating sound waves are observed (Everest and Pohlmann 2001). The sound radiates outward spherically and the area of the sphere equals $4\pi r^2$.

The intensity of a point source is therefore inversely proportional to the square of the distance from the source:

$$I = \frac{W}{4\pi r^2} \tag{2.5}$$

where W denotes the power of the source, I denotes the intensity of sound per unit area, and r denotes the radial distance from the source. Since W and 4π are constant for an individual source, the intensity reduces by a factor of 4 when the distance from the source doubles and the intensity reduces by a factor of 9 when the distance triples.

Due to the wide range of sound pressures and intensities in real acoustic environments, they are often expressed in logarithmic scales. Intensities which can be perceived by humans range from approx. 10^{-12} to $10\ W/m^2$. The most common logarithmic scale for defining sound level is the decibel (dB) scale. The intensity becomes the Intensity Level (IL) and is expressed as:

$$IL = 10\log\frac{I}{I_{ref}}(dB) \tag{2.6}$$

where I_{ref} denotes reference intensity and equals $10^{-12}\ W/m^2$. If we replace intensities with pressures, we get the definition of Sound Pressure Level (SPL):

$$SPL = 20\log\frac{p_e}{p_{ref}}(dB) \tag{2.7}$$

where p_e denotes the amplitude of measured effective pressure and p_{ref} denotes the reference pressure, which equals $20\mu Pa$. Both I_{ref} and p_{ref} correspond to the threshold of human hearing.

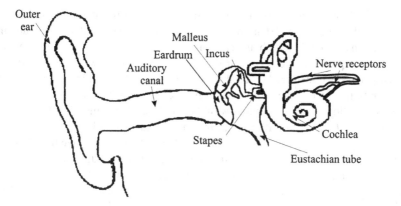

Fig. 2.1 Basic structure of the human ear (Sodnik 2007)

Since the intensity of sound is very difficult to measure, pressure and SPL are the most accessible parameters to measure in acoustics.

2.2 Sound Perception and Psychoacoustics

The study of the perception of sound by humans combines the knowledge and scientific methods of psychology and psychoacoustics. Psychoacoustics studies the physical structure of the human ear, sound propagation channels and their transfer functions, as well as human subjective perception of sound and its interpretation. The basic function of the human hearing system is to convert physical acoustic waves of sound pressure into the electrical discharges that are sent to the brain and which create the sensation called sound.

2.2.1 Human Ear

The human hearing organ consists of three main parts: outer ear, middle ear and inner ear (Fig. 2.1) (Allen and Neely 2006; Everest and Pohlmann 2001). Each of these three parts plays an important role in the process of perception and interpretation of sound. The outer ear consists of a pinna and an ear canal which ranges from 2 to 3 cm in length. Their primary role is to gather as much sound energy as possible and focus it in the eardrum, which is the outer part of the middle ear. The pinna acts as an antenna which intercepts sound waves and filters them according to their individual shape and size. It also serves as the external protection of the ear canal and the ear drum. The ear canal, with its specific length, acts as an amplifier which significantly boosts the frequencies from 3 to 6 kHz. These frequencies are very important for human speech and speech communication.

The middle ear consists of an eardrum, and three tiny interconnected bones. These are called the malleus (hammer), the incus (anvil), and the stapes (stirrup) and are located in an area field with air. The eardrum acts as an elastic membrane which vibrates when pressed by external air pressure. The vibration with the corresponding frequencies is transferred to the three small bones which again act as an amplifier of the transmitted wave. The stapes is physically connected to the inner ear and transfers vibrations to the oval window of the cochlea, which amplifies the vibration by a factor of 15. On the other hand, the middle ear also allows the impedance matching of a sound wave in the air (low impedance) to a sound wave traveling in a system of fluids and membranes in the inner ear (high impedance). Without this matching, the majority of sound waves would simply be reflected off the eardrum back to the ear canal and out of the ear. Since the middle ear is hollow, there is a significant pressure difference between the middle ear and the outside environment when a human moves to a low-altitude environment. The pressure is equalized through the Eustachian tube, which links the middle ear to the nasopharynx (the uppermost part of the throat). The tube is usually closed and opens when yawning or swallowing in order to equalize the difference in pressure.

The inner ear consists of a cochlea, semicircular canals, and detection cells (hair cells). The cochlea is named for its shape (e.g. spiral-shaped cavity) and reaches approx. 3 cm in length if extended. It is filled with a special fluid which acts as a kind of accelerometer and is responsible for maintaining the balance of the entire human body. The most important part of the cochlea is the organ of Corti, which is provided with auditory sensory cells and responds to fluid-borne vibrations. The mechanical vibrations of the oval window are transformed into hydraulic pressure waves with different amplitudes. The exact position of the maximum amplitude of the specific wave corresponds to the characteristic frequency of that wave. Higher frequencies result in higher amplitudes closer to the oval window, while lower frequencies result in higher frequencies close to the end of the cochlea corridor. By the time the sound waves reach the organ of Corti their pressure amplitude is approx. 22 times lower than the amplitude of air entering the ear canal. There are between 15,000 and 20,000 auditory nerve receptors in the Corti capable of activating electric signaling to the auditory cortex in brain.

Humans perceive sounds with frequencies ranging from 20 Hz to 20 kHz, but the upper boundary decreases significantly with age. The majority of human speech communication ranges from 200 Hz to 8 kHz. Sound with frequencies below 16 Hz is called infrasound and it is mostly produced by large machines at construction yards, large instruments, and earthquakes. The absorption of infrasound in a free field is very low due to the very low frequencies, which results in ranges of sound propagation up to 10,000 km. Humans perceive infrasound as external air pressure in the stomach, lungs and chest. Sound with frequencies from 20 kHz up to 250 MHz is called ultrasound. It is mostly produced by machines with fast-spinning parts, rocket engines, and different ultrasonic devices. Certain animals, such as dolphins and bats, are able to produce and perceive ultrasound and use it for communication. Humans can sometimes perceive ultrasound as a special form of pain in the ears.

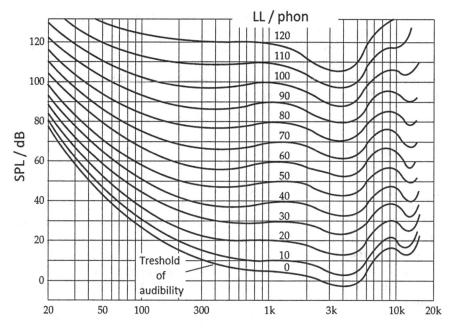

Fig. 2.2 Equal-loudness contours of the human auditory system for pure tones (Everest and Pohlmann 2001)

2.2.2 Loudness and Pitch

In the previous chapter we defined IL and SPL as two physical quantities defining the amplitude of sound pressure or energy on a logarithmic scale. The reference values in the definition of IL and SPL correspond to the threshold of audibility at 1 kHz, since the threshold of audibility is frequency dependent. This nonlinearity corresponds to the nonlinear sensitivity of the human ear. The subjective perception of SPL by humans is defined as Loudness Level (LL) and it is measured in phons. 1 phon corresponds to 1 dB of SPL at 1 kHz. The dependency of sensitivity on frequency is expressed as a form of equal-loudness contours, which have been empirically measured by Fletcher and Munson (1933).

These contours reveal the lack of sensitivity of the human hearing system to low frequencies (bass tones) and the high sensitivity to frequencies between 3 kHz and 6 kHz (Fig. 2.2). The latter correspond to the audible range of human speech and the human vocal system.

A subjective term related to the perception of frequency by humans is called pitch (Everest and Pohlmann 2001). It is a function of frequency, but its relation is non-linear. Frequency is a physical quantity measured in Hertz, while pitch is commonly expressed in mels. The subjective perception of frequency also depends on the current SPL of the sound. A reference pitch of 1000 mels is therefore defined as the pitch of a 1000 Hz signal with an SPL of 60 dB.

Listening experiments have revealed that the human auditory system can detect approx. 300 different intensity levels and approx. 1400 different pitches. That gives us approx. $300 \times 1400 = 420,000$ detectable combinations in total. These numbers relate to very specific situations when only single-frequency sounds were used in a listening test. In the case of more complex sounds, the threshold can drop down to only approx. 50 pitch–loudness combinations.

2.2.3 Masking

Multiple simultaneous sound sources can interfere with one another in different ways. One sound can, for example, be inaudible due to a noise or another louder sound source. The effect is called masking and the masked threshold is the lowest SPL of the perceived sound when being played simultaneously with a specific masking noise (Gelfand and Levit 1998). The final extent of the masking effect depends on the characteristics of the target sound, the masker, and also of the individual specifics of the listener. A typical example is a masking effect between two sounds of similar frequencies and different loudness where the sound source with lower loudness becomes inaudible or the two separate sounds are heard as one combination tone. The ability to hear both frequencies and to hear them separately depends on their position within the critical bands (Boer and Bouwmeester 1974)—auditory filters created by the cochlea. If both frequencies lie in the same critical band, the human auditory system will be unable to separate them and a masking effect will occur. The masking pattern changes depending on the frequency of the masker and its intensity. If two simultaneous sounds are audible but the listener is unable to distinguish them and to define which elements of the perceived signals correspond to each of them, the effect is called informational masking (Gelfand 2010).

Sometimes a sound can also become inaudible due to another sound if they are not played simultaneously but one after the other (Moore 2004). This effect is called temporal masking and can affect sounds following the masker (post-masking) or even preceding it (pre-masking). Both temporal masking effects attenuate exponentially, with the pre-masking effect lasting up to 20 ms and the post-masking effect up to 100 ms.

The time required for perceptual analysis of a sound is approximately 250 ms (Demany and Semal 2007) and depends on pre-perceptual auditory memory (PPAM). Therefore, the duration of the pause between two consecutive sounds affects their perception. When the first sound is played, it is stored in PPAM and if the second one is played in less than 250 ms after the first one the perceptual analysis of the first sound is interrupted. Its perception and interpretation is therefore limited and incomplete.

Fig. 2.3 Spherical
coordinate system for
describing location of
sound sources relative to
the user (Sodnik 2007)

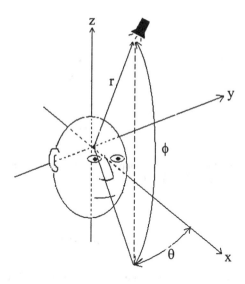

2.3 Localization of Spatial Sound

The human auditory system is capable of determining the exact location of the origin of a sound source. This process is called localization and it is of vital importance for orientation in space, communication with multiple people simultaneously, driving a vehicle or bicycle, etc. When referring to the position of the selected sound source in space we usually define its position relative to the listener. The spherical coordinate system is the most appropriate for defining this relation by selecting the listener's head as its origin (Fig. 2.3).

We define the angle θ in the horizontal plane as the azimuth of the sound source and the angle $\varnothing$ in the vertical plane as the elevation. r refers to the radial distance from the center of the head. The ranges of the individual quantities are as follows:

$$-180° \leq \theta < 180°, -90° \leq \varnothing < 90°, 0 \leq r < \infty \tag{2.8}$$

The determination of the location of a sound source starts at the pinna (Everest and Pohlmann 2001). A large number of different rays of sound combine at the entrance to the auditory canal. Different rays correspond to different propagation paths summing unreflected direct sound and sound reflected from various surfaces and body parts. This combination encodes the directional information of an individual sound source and can be described as a multiplicity of sound rays coming from a specific azimuth θ and elevation $\varnothing$ relative to the human listener.

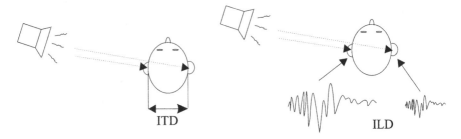

Fig. 2.4 The time and amplitude difference between incident waveforms for the direct and shaded ears (Sodnik et al. 2005)

2.3.1 Inter-aural Time and Level Difference

One of the oldest theories in the field of sound localization is called the duplex theory, which defines the Inter-aural Time Difference (ITD) and Inter-aural Level Difference (ILD) as the two major binaural cues for defining the location of the selected sound source (Strutt 1907) (Fig. 2.4).

ITD refers to the delay between the impacts of a sound wave in the left and right ear, which corresponds to the difference of the distance between the ears and the sound source. The maximum possible binaural delay is approx. 0.65 ms (Rumsey 2001). It is also important to distinguish between the binaural delay caused by a single sound source and the delay that occurs between the ears in the case of two almost simultaneous sound sources at different locations. The latter causes the brain to perceive just one single source from the location nearer to the earliest of the two sources.

ILD, on the other hand, refers to the amplitude difference of sound waves when reaching the left or right ear. The amplitude changes due to multiple reflections from various obstacles on the acoustic path. The wavelength of sounds with the frequencies of approx. 1500 Hz becomes comparable to the diameter of the human head and ITD cues become ambiguous (Cheng and Wakefield 1999). The ear farther away from the sound source get shadowed by the head for all frequencies above 1500 Hz and consequentially less energy arrives at the shadowed ear than at the non-shadowed ear. In these cases ILD becomes very important for correct perception of the azimuth.

The impact of the two selected factors has been shown and defined through various listening experiments. Kistler and Wightman (1992) systemically manipulated the ILD and ITD of virtual sound sources and concluded that ITD is of vital importance when localizing sources of low frequencies, while on the other hand ILD is more important when dealing with sound sources of high frequencies (above 5 kHz). The duplex theory is undoubtedly attractive and relatively simple to understand, but it only explains perception of the azimuth or left–right direction (Macpherson and Middlebrooks 2002). It was later shown that localization of sound sources cannot depend solely on binaural cues since relatively accurate localization in median plane ($\theta = 0°$; $90° \leq \varnothing < 90°$) is also possible. In this case ITD and ILD are minimal or close to zero. The latter indicates the importance of some additional factors and monaural cues for sound localization. Nowadays it is generally established that the shape and the size of the pinnae, head, shoulders, and other parts of the human body are of key importance for accurate sound localization. They act as individually-tailored filters which differ from person to person and specifically change the frequency spectrum of all sound signals.

2.3.2   Frequency Response of Human Auditory Channel

The propagation of a sound wave in free space can be described as a time-independent linear system. A response, $y(t)$, of such a system is usually defined as the convolution of an input signal, $x(t)$, and of a unit impulse response, $h(t)$ (Oppenheim et al. 1989):

$$y(t) = x(t) * h(t) \tag{2.9}$$

A system response can also be defined in the frequency domain:

$$Y(\omega) = X(\omega) H(\omega) \tag{2.10}$$

where $Y(\omega)$ and $X(\omega)$ denote Fourier transforms or frequency responses of output and input signals, respectively, and $H(\omega)$ denotes the transfer function of the system. In the case of the human auditory system, $H(\omega)$ describes the acoustic path between a sound source and an eardrum in the selected human ear, which includes the impact of various body parts.

The significant importance of the frequency response of the human auditory system for localization has been proven by various localization experiments with just one ear (Wightman and Kistler 1997) or by manipulating the frequency spectrum played to an individual ear (Langendijk and Bronkhorst 2002). These experiments revealed the importance of different frequency bands for correct localization and proved that a frequency band between 5.7 kHz and 11.3 kHz is vital for correct localization of the elevation of a sound source. On the other hand, a frequency band between 8 kHz and 16 kHz proved to be of vital importance for determining whether a sound source is located in front of or behind a listener. Besides these broad directional bands, there are also various narrow peaks and notches in the frequency spectrum corresponding to certain locations.

Another experiment also demonstrated the importance of lower frequencies for correct elevation localization (Algazi et al. 2001a). This part of the frequency spectrum is particularly influenced by reflections of sound waves from larger objects, such as the head, torso, arms, legs, etc. The latter causes changes in ITD and ILD

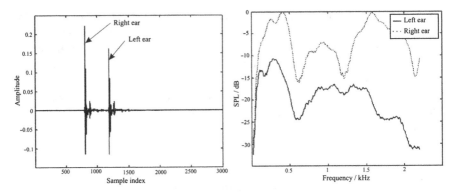

Fig. 2.5 HRIRs (*left side*) and HRTFs (*right side*) of both ears measured at an azimuth of 25° and an elevation of 0° (Sodnik 2007)

for different elevations and improves the accuracy of perception. It also depends on whether a person is seated or standing, and still or in motion. Small head movements also contribute significantly to the accuracy of localization. By tilting the head slightly left or right, and thus placing the ears at different heights, some kind of inter-aural cues can also be produced for vertical localization. However, this effect can only be produced when listening to real sources or real speakers and not in virtual conditions when using headphones.

2.3.2.1 Head-Related Transfer Functions

Individual body parts, such as the head, torso and pinnae cause specific and characteristic changes in the frequency response which depend on the individual properties of a person. Due to this relation a transfer function $H(\omega)$ for each ear is called Head Related Transfer Function (HRTF) (Cheng and Wakefield 1999). A single HRTF therefore represents an individual's left or right ear far-field frequency response as measured from a specific point in the free field to a specific point in the ear canal. In fact HRTFs for both ears are usually measured as Head Related Impulse Responses (HRIRs) for arbitrary spatial positions (Fig. 2.5). The procedure will be described in one of the following chapters.

The shape and the size of every human pinna are unique, so pinnae are like fingerprints for each individual person (Rumsey 2001). HRTFs are, therefore, also unique to each listener and almost impossible to generalize across the larger population. Listening experiments with people who were given another person's HRTFs revealed a significant reduction of localization accuracy. After some time they were able to slightly adapt to new localization cues and increase their performance.

2.3.3 Loudness and Duration

Localization accuracy also depends on the loudness and duration of sound signals (Macpherson and Middlebrooks 2003; Vliegen and Opstal 2004). The accuracy improves linearly with the duration of the sound signals, but only until approx. 30 ms. No additional improvement has been shown when increasing the duration beyond this limit. Similarly, localization accuracy improves rapidly with increased loudness, but again only to approx. 55–65 dB SPL. For higher levels of SPL, the accuracy drops rapidly.

2.3.4 Perception of Distance

The filtering of pinnae and HRTFs do not play an important role in detecting the distance of an arbitrary sound source. The four most important acoustic factors for the perception of distance are intensity of sound signal, ratio between the direct and the reflected sound, spectrum of sound signal, and ITD (Begault 1991a; Zahorik 2002a, b).

As stated in Eq. 2.5, in ideal conditions the intensity of a sound source drops linearly with the square of the distance. Doubling the distance corresponds to a decrease in SPL by 6 dB. This factor could be problematic if the original SPL of sound source changes with time.

A certain amount of sound waves from a specific source travels directly and unobstructed to human eardrums while certain sound waves reflect from various obstacles and surfaces in space and lose their power and intensity. This ratio of direct to reverberant sound is directly related to the source distance, particularly in enclosed spaces where multiple reflections occur (Rumsey 2001). The reverberation time and the early reflection time enable the brain to acquire information on the size of the surrounding space and the distances to various surfaces and sound sources.

The air as a typical medium for propagation of sound also causes some characteristic changes in a sound wave. It attenuates high frequencies for several dB in every 100 m. The sound spectrum is also affected and characteristically reshaped due to various reflections in the acoustic path. Due to these two factors a sound source at longer distances sounds more reverberant and with lower pitch.

2.3.5 Spatial Resolution of the Human Auditory System

By spatial resolution or Minimum Audible Angle (MAA) we refer to the capability of the human listener to distinguish the difference between two individual sound sources in near proximity. MAA defines the minimal spatial angle between the two sound sources at which they are still perceived as individual sources.

For smaller angles, the sources are perceived as just one single point source. This limitation is of vital importance for the design of spatial auditory interfaces, since it determines the maximal available resolution, similar to the resolution of computer displays and other output devices.

MAA has been determined and measured with different experiments, including real sound sources played through external speakers and virtual sources played through headphones (Sodnik et al. 2005). The final measured resolution proved to be significantly different for the azimuth and for the elevation. There is also a significant difference in resolution between real and virtual sound sources. An average resolution in the horizontal direction was shown to be approx. 2° for real sources and approx. 4.1° for virtual sources. For the vertical direction, the resolution decreases to approx. 6° for real sources and to 23.5° for virtual sources.

2.3.6 *Auditory Versus Visual Perception*

Some basic limitations of human auditory perception can also be illustrated by comparing the latter to human visual perception and the sense of sight. The visible angle of the human eyes is approximately 80° in the horizontal plane and 60° in the vertical (Perrott et al. 1991). The area of sharp vision or high focus is only approximately 2° around the central area of this window and decreases from the center to the peripheral areas. Sound, on the other hand, is omnidirectional and has no such limitations in perception.

The most important difference between auditory and visual perception is in their information processing ability. The comparison between two perception mechanisms can be illustrated as a comparison between two communication channels. The visual channel can be represented as a parallel channel, while the auditory channel acts as a serial channel. The reported bitrate for a visual channel presented with a 1024×1024 pixel bitmap image with 8-bit color depth is approx. $4.32 \cdot 10^6$ bits / s (Koch et al. 2006). In the visual channel a great amount of information is perceived simultaneously. The auditory channel, on the other hand, is much narrower with a bandwidth of only approx. 9.900 bits/s (Jacobson 1951), resulting in much lower processing capabilities.

Visual information usually stays on the screen for a longer period of time and is also processed in parallel, while perception of sequentially delivered auditory information also depends on the above-mentioned PPAM. It sometimes has to be replayed several times in order to be remembered and perceived correctly.

Since people are primarily visually oriented, their sense of sight is largely occupied most of the time. It is therefore often insensitive to details and minor changes in the perceived image. Sound, on the other hand, can draw our attention at any time and is perfect for delivering temporally important information. The response time of the human auditory system is also faster than the response time to visual stimuli (Nees and Walker 2009).

2.3.7   Attention and Distraction

According to Anderson (1990), attention is the behavioral and cognitive process of selectively concentrating on one aspect of the environment while ignoring other things, whereas distraction is the divided attention of an individual or group from the chosen object of attention toward the source of distraction. Humans only have a limited amount of attention at a certain times and their ongoing tasks can either use the same attention resources or different ones (Navon and Gopher 1979; Wickens 1984). When the same resource is used in two or even more simultaneous tasks they interfere with each other and their performance drops. A typical example of sharing visual resources in different tasks is when operating a mobile phone or other communication device through a visual interface while walking, cycling, driving, etc. These are considered primary tasks and the use of a mobile phone causes distraction from these primary tasks. On the other hand, the utilization of the same device through an auditory interface would not compete for visual attention and would result in lower distraction from the primary task.

The inter-distraction of multiple simultaneous tasks and the person's ability to pay attention to one selected task also depends on their complexity (Iqbal et al. 2010; Lee et al. 2001; McKnight and McKnight 1993; Ranney et al. 2000). Studies with drivers have shown that physical and cognitive distraction can also affect visual attention, resulting in slower reaction times and decision-making, decreased performance in visual search of patterns, and decreased driving performance in general (Young et al. 2003). Regardless, attending to visual and auditory stimuli simultaneously is always easier than attending to two visual or two auditory channels (Hirst and Kalmar 1987).

One additional aspect of auditory attention is a listener's ability to detect an individual sound signal from a background noise or from a stream of multiplexed signals (Hafter et al. 2007). An example is the ability to concentrate and to listen to a single talker in a multi-talker environment and to filter out the rest of the simultaneous sounds. This phenomenon is called the "cocktail party effect" (Cherry 1953; Stifelman 1994) and its final affect depends on the distance of the talkers, the SPL of their speaking, the individual characteristics of their voices, gender, etc.

2.4   Generation of Spatial Sound

The use of spatial sound in various applications and auditory interfaces requires its generation and artificial creation of localization cues, as described in previous subchapters. In the process of sound synthesis, all propagation paths from the source at the desired position in space to human ears at another position have to be calculated and applied correctly. They describe the geometric transformations of the sound sources as they correlate to the acoustic environment. These calculations are rather complex and require considerations of the physical dimensions of the environment, the exact positions of sources and listeners, as well as other objects and obstacles in the acoustic path. In the previous chapter, we also described the importance of the frequency response of the human auditory system and its role in localization performance. Another more feasible and commonly-used approach is to take a binaural recording of the acoustic path and acquire its transfer functions. This procedure is described in the following chapters.

2.4.1 Measurement of Head-Related Impulse Responses (HRIRs)

In order to generate accurate and authentic spatial sound, transfer functions of the human auditory system have to be measured, including the individual properties of the sound source, the acoustic path, the shape of the head, torso and pinnae, and the acoustic equipment used in the measurement. Transfer functions of such complex systems are commonly measured as impulse responses for each ear separately. In the case of the human auditory system, these impulse responses are called HRIRs.

An example of the measurement of personalized or individualized HRIRs was described by Sodnik (2007). The experiments took place in an anechoic and quiet room in order to avoid reflections from any unwanted obstacles in the acoustic path and also to avoid any unwanted noise. A listener or a test subject had to sit still in the center of the room. Small microphones were inserted into the entrance of the ear canal. Ideally the microphones should be positioned as close as possible to the eardrums, yet this is impossible with human test subjects. It has been shown that measurements made at the entrance of the ear canal contain all the directional information of the individual listener (Moller et al. 1995; Rumsey 2001). The ear canal response has no directional encoding function. The center of the user's head represented the center of the virtual coordinate system (Fig. 2.3), which was also indicated with red tape on the floor (see Fig. 2.6). The coordinate system was used to define the relative spatial position of the sound source and the user.

The speaker was then placed in an arbitrary spatial position relative to the user and an impulse response was measured for that spatial position. The speaker was then moved to a new azimuth angle or lifted to a new elevation and the measurement was repeated. In fact, this measurement had to be repeated for all desired spatial positions. The procedure can be optimized and sped up by mounting a speaker on a robotic arm which can move quickly to any spatial position within its range. The alternative is also to have a speaker at a fixed position and to rotate and lift the listener's chair accordingly.

A unit impulse is a theoretical signal and cannot be used in practical measurements. Instead, short sequences of pseudo white noise with a flat frequency spectrum were used as test signals. This test signal of length N (number of samples) was then played from a speaker and simultaneously recorded with the two microphones inserted in the listener's ears. The transfer function of the system or the HRTF of an

Fig. 2.6 Anechoic room
setup for the measurement
of HRIRs (Sodnik 2007)

individual ear could then be calculated as a ratio between the Fourier transforms of
played (H_{inp}) and recorded (H_{out}) signals:

$$HRTF_{left} = \frac{H_{inp}}{H_{out\_left}} \quad HRTF_{right} = \frac{H_{inp}}{H_{out\_right}} \tag{2.11}$$

Impulse responses or HRIRs were then acquired by the inverse Fourier transform
of the calculated HRFTs. Similarly, impulse responses of acoustic equipment can
also be measured in order to be compensated for in the final HRIR measurements.
An example of pairs of HRIRs and HRTFs is shown in Fig. 2.5.

Technically speaking, HRTFs represent minimum-phase Finite Impulse
Response (FIR) filters for a specific spatial position, which also include information
on ITD and ILD (Cheng and Wakefield 1999). ITD or time delays are encoded into
the filters' phase spectrum, while ILD relates to the overall power of the filters.
HRTFs can also be represented as minimum-phase systems allowing us to separate
ITD information from the FIR specification. Minimum-phase filters have a minimum group delay property and minimum energy delay property and most of the
energy occurs at the beginning of their impulse response. In the case of HRTFs, left
and right minimum-phase filters both have zero delay. Consequently, the complete
information on a specific spatial location consists of only left and right magnitude
responses and the corresponding ITD.

HRIRs of the left and right ear can now be used as FIR filters of length N to manipulate various sound sources to be virtually played from the spatial position where original HRIRs were measured. An arbitrary mono sound signal can be enriched with spatial properties by calculating the convolution between the two filters and the signal itself:

$$x_{left,\theta,\varnothing}[n] = \sum_{n=0}^{N-1} x[n-k] hrir_{left,\theta,\varnothing}[k] \tag{2.12}$$

$$x_{right,\theta,\varnothing}[n] = \sum_{n=0}^{N-1} x[n-k] hrir_{right,\theta,\varnothing}[k] \tag{2.13}$$

where x denotes the selected mono sound signal, $hrir_{left,\theta,\varnothing}$ and $hrir_{right,\theta,\varnothing}$ denote the appropriate impulse responses, and the calculated $x_{left,\theta,\varnothing}$ and $x_{right,\theta,\varnothing}$ denote the target signals representing spatial sound x at azimuth θ and elevation $\varnothing$.

The procedure described above requires measurements for each person and for each spatial position. Individualized HRIRs are very accurate and only enable correct localization for the person participating in the measurement. For any other person the error is approximately similar if using another person's HRIRs or a set measured with artificial heads. It has been found that some people localize sound much better than the others and their individual HRIRs perform better for a wider range of people (Begault 1991b; Rumsey 2001). It has also been proven that people can gradually adapt to a new set of HRIRs and increase localization performance over time.

2.4.1.1 Dummy Heads

Measurements of HRIRs can be very inconvenient for a person involved as a test subject due to the long duration of the test and the unpleasantness of putting micro-phones into one's ear canals. Unwanted movements and noises of the human test subjects can also affect the final results. HRIR measurements are therefore often simplified and done with mannequins and dummy heads (Fig. 2.7). These are models of human heads with microphones built into the ears. Some models of dummy heads include shoulders or complete torsos since reflection from shoulders also contributes to the shape of the final HRIRs. The ears of these dummy heads can often be changed and therefore vary in shape and size.

Measurements with dummy heads result in non-individualized or generalized HRIRs which are less accurate and result in poorer localization performance. On the other hand, they can be recorded much faster and for a larger number of spatial positions.

2.4.1.2 Equalization

Recording and reproduction of sound involves different equipment with its own frequency response, which is also reflected in the final output (White and Louie 2005). In the case of HRIR measurements, the goal is to solely measure the

Fig. 2.7 Neumann KU
Dummy Head Microphone
(Performance Audio 2014)

frequency response of the acoustic path between a speaker and the entrance to the ear canal. The response should therefore not reflect the frequency characteristics of the speaker, microphones, sound cards, amplifiers, etc. A process called equalization is used to alter these responses with special linear filters in order to make them more or less flat. The filters represent antiphase versions of their own frequency responses.

In the reproduction process the headphones should also always be equalized. They should have flat response at the point in the ear where the binaural recording microphones were originally inserted. Sometimes we want to emulate the free-field response, which also includes the timbre of sounds produced by the loudspeakers. In the process of headphone equalization the timbre can also be artificially added to their frequency response. On the other hand, such equalization affects the directional information in the reproduced spatial sound and decreases the localization accuracy. Therefore, there is always an inevitable tradeoff between accurate timbre matching and accurate localization.

2.4.2 Interpolation of HRIRs

With real test subjects or with dummy heads, HRIRs are always measured at a finite number of points in space. In order to provide a continuous source-position domain, different interpolation techniques can be used in a time or frequency domain. One of the simplest procedures is a time-based linear interpolation (Sodnik et al. 2005), which has proved to have the largest signal-to-deviation ratio. A simple linear interpolation is the most appropriate method for interpolating functions where azimuth spacing is less than $10°$. The first step of interpolation requires two impulse responses with neighboring azimuths θ_1 and θ_2 and common elevation $\varnothing$ to be aligned

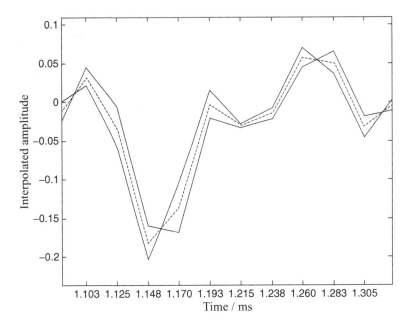

Fig. 2.8 Linear interpolation of intermediate impulse responses (Sodnik et al. 2005)

on the time axis. The latter requires the determination of their relative delay. Since impulse responses of all neighboring HRIR's are very similar, the delay can simply be extracted as the position of the maximum of the cross-correlation function:

$$R_{\theta_1,\theta_2,\varnothing}[m] = \frac{1}{N}\sum_{n=1}^{N-1} hrir_{\theta_1,\varnothing}[n] \, hrir_{\theta_2,\varnothing}[n+m] \qquad (2.14)$$

where $hrir_{\theta_1,\varnothing}$ and $hrir_{\theta_2,\varnothing}$ denote the two neighboring HRIRs and N denotes the number of samples in both impulse responses. The delay Δs expressed in the number of samples can now be extracted as the position of the maximum of Eq. 2.14. The new interpolated HRIR is now equal to:

$$hrir_{\frac{(\theta_1+\theta_2)}{2},\varnothing}\left[n+\frac{\Delta s}{2}\right] = \frac{hrir_{\theta_1,\varnothing}[n] + hrir_{\theta_2,\varnothing}[n+\Delta s]}{2} \qquad (2.15)$$

A new HRIR with the azimuth at the center of any two neighboring HRIRs can be interpolated with linear interpolation (Fig. 2.8). Similarly HRIR's for unmeasured elevation angles could also be gathered with a similar procedure.

2.4.3   Public HRIR Libraries

Numerous free and publicly-available HRIR libraries can be found on the internet, the majority of which comprise measurements for more than 500 spatial positions and for many test subjects. The two most popular and widely-used libraries in academia are the MIT Media LAB HRIR library (Gardner and Martin 1994) and the CIPIC library recorded at U.C. Davis CIPC Interface Lab (Algazi et al. 2001b).

The MIT library consists of measurements for 710 spatial positions covering azimuths from 0° to 360° with a resolution of 5° and covering elevations from −40° to 90° with a resolution of 10°. The measurements were acquired with the artificial mannequin called Kemar. For each position there is a pair of FIR filters for each individual ear, consisting of 512 samples in all.

The CIPIC library consists of measurements for 1250 spatial positions and for 45 test subjects, including artificial heads and real test subjects. In this case, the spatial resolution is 5° for the azimuth and elevation. The length of FIR filters is 200 samples.

2.4.4   Modeling HRTFs

The majority of HRIR libraries contain a set of FIR filters systematically arranged to cover multiple spatial locations. Let us consider a set for 1500 spatial positions which contain 3000 filters (for both ears), each consisting of 1024 coefficients. In total, such a library contains $3000 \times 1024 = 3,072,000$ coefficients. Reproduction of the spatial sound in real time with the aid of these filters requires enormous computational power and memory. Various studies have focused on finding a simple and more efficient definition of HRTFs. Infinite Impulse Response (IIR) filters represent a promising solution, as they enable specification of transfer functions with significantly fewer coefficients. Long impulse responses describing various properties of an acoustic path can be defined with only a few poles. A combination of zeros (Haneda et al. 1999) and poles (Kulkarni and Colburn 2004), or just poles, in an IIR filter can accurately define system response for a specific spatial position. Another approach is to model HRTFs with a set of resonators that represent recursive filters with a complex conjugate pair of poles on the unit circle (Susnik et al. 2003). This model consists of six resonators, a notch filter, a delay unit, and an amplifier. The central frequencies of the resonators and the notch filter have been set empirically by visually fitting an original and modeled HRTF.

The compression of data with the aid of Principal Component Analysis (PCA) is also feasible and represents an interesting alternative to the FIR representation of HRTFs (Kistler and Wightman 1992; Sodnik et al. 2006). It is based on the presumption that there is a significant amount of redundancy in HRTFs when comparing responses for different azimuths. Subjective tests confirmed that a set of 37 HRTFs can be accurately represented with only a few PCA coefficients and their corresponding weights (Sodnik et al. 2006).

2.5   Reproduction and Rendering of Spatial Sound

The most common device for sound reproduction is a loudspeaker. It is an electro-acoustic transducer which produces sound based on an electrical input signal. In the case of spatial sound, at least two speakers are required in order to reproduce binaural differences and artifacts.

2.5.1   Headphones

Reproduction based on headphones is significantly different from any external loudspeaker configuration since it can directly feed each ear with an individual sound signal or channel. The two signals contain the correct ITD, ILD and inter-aural spectral differences. HRTFs and HRIRs can therefore be directly applied for the reproduction of spatial sound through headphones. Any mono signal can simply be filtered with the corresponding pair of HRIRs and played from a virtual spatial position.

However, there are still several important problems related to the correct and authentic reproduction of spatial sound through headphones (Rumsey 2001). We have already described the problem of the individuality of HRIRs, which prevents correct generalization for a wide range of listeners or results in a significant decrease of reproduction quality.

The other problem is a lack of head movements, which help to resolve directional confusion in real-life listening especially front–back confusions. Head movements can be simulated by using a head tracker which provides real-time information on current head position and orientation. This information enables modification and adaptation of binaural cues in current HRIRs emulating realistic changes in signals sent to the ears when moving the head. Such real-time adaption and calculations of new responses requires large processing power, as the final latency cannot exceed 85 ms.

There is also no visual feedback on the position of the loudspeakers, which can help to determine the correct position of the sound source. This can only be resolved in a 3D virtual reality environment which incorporates visual and audio information and provides visual cues on the position of loudspeakers.

2.5.2   Binaural Signals on Loudspeakers

Sometimes we want to play binaurally-recorded signals on loudspeakers. They are placed at a certain distance from the listener and the crosstalk effect occurs (Rumsey 2001). Each ear receives signals from all loudspeakers and it is impossible to play only one channel to one selected ear. In the case of stereo loudspeakers, the left ear,

Fig. 2.9 Conventional
crosstalk cancellation for
stereo loudspeakers
(Rumsey 2001)

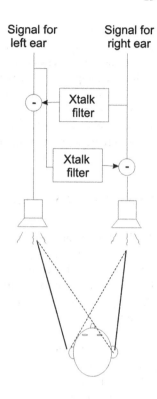

for example, receives the left channel first, since the left loudspeaker is usually
physically closer. Just a fraction of time later, it also receives the right channel or the
sound signal from the right loudspeaker. Both received sound signals are naturally
filtered with HRTFs corresponding to the positions of the loudspeakers relative to
the listener. This results in inaccurate binaural cues at the listener's ears and prevents correct spatial sound reproduction.

With some additional processing, it is possible to preserve full 3D perception of
loudspeakers by correcting or removing the crosstalk effect. The procedure is called
crosstalk canceling or transaural processing (Jo et al. 2011). The idea is to the feed
the left ear with only the left channel and the right ear with the right channel. This
is possible by also feeding the left ear with the antiphase version of the right channel
and the right ear with the antiphase version of the left channel. Both antiphase signals have to be filtered and delayed according to the HRTF characteristic of the
crosstalk acoustic path (Fig. 2.9).

Crosstalk cancelation enables authentic perception of sound sources at any spatial position. Although only two speakers placed in front of the listener are used for
reproduction, the sound can be perceived as coming from behind the listener. The
major limitation of this approach is a rather small hot spot of accurate perception.
A small displacement can cause completely inaccurate perception and the spatial

effect disappears. Due to the small hot spot, it is almost impossible to support reproduction for several listeners simultaneously and to achieve a certain degree of robustness in such a system.

Some examples of successful and accurate solutions with crosstalk cancelation are 3D audio systems for desktop computers or home theater systems. Listeners are almost stationary in these environments and their position is well known. The designers and manufacturers of these systems must find a tradeoff between localization accuracy, support for multiple listeners, and robustness.

2.5.3 Virtual Surround Systems

HRTFs and binaural processing techniques can also be used to create a virtual surround system where only two loudspeakers can simulate five or even more virtual loudspeakers. In the case of a virtual surround system that consists of five speakers, three speakers have to be virtualized (C—central; RR—rear right; RL—rear left). The biggest challenge is to virtualize two rear speakers which have to produce sound from behind the listener. All virtual speakers have to be binaurally processed with the corresponding HRTFs, usually one for the central position ($\theta_c = 0°$), and the two rear positions ($\theta_{rr} = 110°$ and $\theta_{rl} = -110°$). These are the conventional positions of speakers in 5.1 surround systems. A special processing unit for crosstalk cancellation eliminates crosstalk effects and mixes transaural virtual channels with the front left and front right channel (Fig. 2.10).

Such systems are very appropriate for environments where real loudspeakers cannot be used for various reasons. They can be implemented as software components in consumer televisions and are optimized for a relatively wide listening area. The localization accuracy of rear virtual speakers drops significantly if a listener moves out of the hotspot. However, the correct localization is usually not that important, as surround channels for the majority of entertainment content are not meant to be very precisely localized.

2.5.4 Virtual Acoustic Environments (VAEs)

Binaural processing and spatial sound rendering require the computationally-intensive synthesis of appropriate frequency responses. In order to implement methods and tools for the virtual positioning of sound sources based on the algorithms described in previous chapters, a synthesis has to be conducted offline or with highly-specialized hardware. Simplified processing techniques that only emphasize a limited number of directional cues and parameters are commonly used as a tradeoff between localization accuracy and computational demands. For real-time reproduction of spatial sound within an application for desktop or mobile computers, it is very convenient to use one of the available virtual acoustic environments (VAEs)

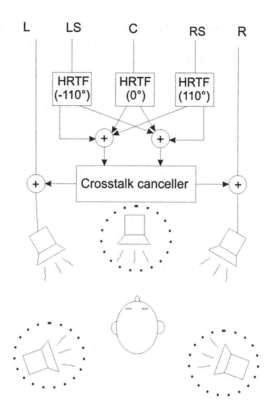

Fig. 2.10 Virtual surround system based on only two stereo loudspeakers and crosstalk cancelation techniques (Rumsey 2001)

(Miller and Wenzel 2002; AstoundSound 2014). Different VAE applications address different aspects of spatial sound synthesis and listening experience and also require different software and hardware rendering systems. They can be utilized as independent applications for sound rendering or integrated into custom software based on predefined Application Programming Interfaces (APIs). In this chapter, we list the basic properties of three widely-used VAEs found on the market, which were also used in our previous research.

2.5.4.1   OpenAL

OpenAL was originally developed by Loki Software (2000) and later maintained by the open source community. It is supported by a variety of platforms, but it is now hosted by Creative Technology (Creative 2014) and therefore proprietary. An open source alternative is also available that it is called OpenAL Soft (OpenAL Soft 2014).

The library represents an audio API for efficient rendering of multichannel 3D positional audio (OpenAL 2014). The programming conventions and APIs resemble OpenGL (OpenGL 2014), which is a well-known and widely-used graphic library for rendering 2D and 3D vector graphics. It supports the generation and reproduction of multiple sound sources at different spatial positions for a single listener, which can also be positioned in space (OpenAL 2014). All spatial locations are expressed as floating-point vectors in a right-handed Cartesian coordinate system (RHS), where in a frontal default view X points right, Y points up and Z points toward the viewer (OpenAL 1.1 2005). A listener L at the origin and two sound sources S_1 and S_2 left and right of the listener would, for example, be defined as:

$$L = [0.0f, 0.0f, 0.0f], S_1 = [-10.0f, 0.0f, 0.0f], S_2 = [10.0f, 0.0f, 0.0f] \quad (2.16)$$

Sources are addressed as source objects that contain information on the position, velocity, direction and the intensity of sound. They operate as pointers to special buffers with corresponding audio data in an 8- or 16-bit Pulse-Code Modulation (PCM) format. The listener object also contains information on position, velocity, orientation and gain. Orientation is defined as two floating-point vectors specifying the "at" and "up" direction. The "at" vector represents the forward direction of the listener, and the orthogonal projection of the "up" vector into the subspace perpendicular to the "at" vector represents the "up" direction. For example, if a user faces forward and away from the camera (in the –Z direction) with the crown of their head facing upward (in the Y direction), the orientation would be defined as:

$$L_{orientation} = \{(0.0f, 0.0f, -1.0f), (0.0f, 1.0f, 0.0f)\} \quad (2.17)$$

The single-listener mode represents the main limitation of the OpenAL library, as it is not appropriate for environments with multiple human participants.

OpenAL is supported by all major platforms, including mobile platforms such as Android, BlackBerry and iOS. It also utilizes an extension mechanism which enables individual vendors to include their own extensions in the original library. In this way they can expose and utilize different features of their proprietary hardware. The OpenAL library supports headphones and various configurations of loudspeakers for the reproduction of spatial sound.

2.5.4.2 Slab3d

Slab3d is another real-time positioning library based on 3D sound processing (Miller and Wenzel 2002; Slab 2014). It was originally developed by the Spatial Auditory Displays Lab at NASA Ames Research Center and was released under a free public license for non-commercial use. Its goal is to provide an experimental platform with low-level control of various signal-processing parameters, such as number, fidelity and positioning of reflections, latency, update rate, etc. It also allows the user to allocate sound sources, define an acoustic scene, configure frame-accurate callbacks and control rendering. HRTF databases are used for binaural processing. It can be utilized through well-managed API for specifying an acoustic scene and setting low-level parameters. It can also take advantage of improvements in hardware performance without extensive software revisions. However, it only supports the Windows operating system from version 2000 on.

Slab3D uses a left-handed Cartesian coordinate system. In this case +X faces to the front through the nose, +Y faces left and +Z faces up through the top of the head. Orientation is defined by yaw (rotation around the X axes), pitch (rotation around the Y axes) and tilt (rotation around the Z axes). It also supports a polar coordinate system where azimuth, elevation and range correspond to the system defined in Fig. 2.3.

Its main advantage is support for numerous input and output formats and devices, such as Waveform-audio devices, 8- or 16-bit PCM disk files, memory-buffered files, recorded inputs, various internet files, etc. Headphones have to be used for reproduction of synthesized spatial sound.

2.5.4.3 SoundScape Renderer

SoundScape Renderer (SSR) is another open-source tool for real-time spatial audio reproduction, currently only available for Linux and MacOS platforms (SoundScape Renderer 2014). It was developed at the Quality and Usability Lab at TU Berlin (2014) and is now being further developed at Institute of Communications Engineering at the University of Rostock (2014).

It provides a variety of rendering algorithms for wave field synthesis, vector-based amplitude panning, dynamic binaural synthesis, binaural room synthesis, etc. (Geier and Spors 2012). The signal-processing core supports multi-threaded processing and the independence of the audio backend. It supports headphones and loudspeaker-based output configurations. In the case of headphones, there is also native support for a number of head trackers. Additionally, any program that produces audio data or any live input from audio hardware can be connected to the SSR and used as a source input.

SSR can be utilized with a built-in GUI or via a propriety network interface. Sound sources with corresponding spatial positions are presented in a spatial audio scene, which can also be saved or loaded from an Extensible Markup Language (XML) file. All SSR rendering modules are available as a software library, which can be used as a stand-alone program or as a plug-in in external audio software. Two important limitations of SSR are the limitation of source positions to the horizontal plane and the possibility of only storing static audio scenes.

2.6 Conclusion

This chapter presented the basic properties of the acoustic wave and its propagation in space. It also explained the functioning of the human auditory system and its limitations, with an emphasis on the localization of spatial sounds. It has been shown that the ability of localization depends on the position of the source relative to the user and the spectral content of the source. The human auditory system uses binaural and monoaural factors to extract the position of any sound source, some of which have to be learned and trained during childhood. Knowledge and understanding of these localization cues enable us to generate and reproduce virtual spatial sound sources by applying prerecorded filters and interpolation algorithms. The complexity and accuracy of the reproduction depends on the available software and hardware.

References

Algazi VR, Avendano C, Duda RO (2001a) Elevation localization and head-related transfer function analysis at low frequency. J Acoust Soc Am 109(3):1110–1122

Algazi VR, Duda RO, Thompson DM, Avendano C (2001b) The CIPIC HRTF Database. In: Proceedings of 2001 IEEE workshop on applications of signal processing to audio and electroacoustics, New Paltz, NY, pp 99–102

Allen JB, Neely ST (2006) Micromechanical models of the Cochlea. Phys Today 45(7):40–47

Anderson JR (1990) Cognitive psychology and its implications. WH Freeman/Times Books/Henry Holt & Co.

AstoundSound (2014) Astound Studios, Llc. http://www.astoundholdings.com/. Accessed 31 Dec 2014

Bamber JC (2004) Speed of sound. In: Hill CR, Bamber JC, Ter Haar GR, editors: Physical Principles of Medical Ultrasound. Hoboken, NJ: Wiley, pp 167–190

Begault DR (1991a) Preferred sound intensity increase for sensation of half distance. Percept Mot Skills 72:1019–1029

Begault DR (1991b) Challenges to the successful implementation of 3-D sound. J Audio Eng Soc 39(11):864–870

Boer ED, Bouwmeester J (1974) Critical bands and sensorineural hearing loss. Int J Audiol 13(3):236–259

Cheng CI, Wakefield GH (1999) Introduction to head-related transfer functions (HRTFs): Representations of HRTFs in time, frequency, and space. In: Audio Engineering Society Convention 107, Audio Engineering Society

Cherry EC (1953) Some experiments on the recognition of speech, with one and with two ears. J Acoust Soc Am 25(5):975–979

Creative (2014) Creative Technology Ltd. http://us.creative.com/. Accessed 31 Dec 2014

Demany L, Semal C (2007) The role of memory in auditory perception. In: Yost WA, Popper AN, Fay RR (eds) Auditory perception of sound sources, vol 29, Springer handbook of auditory research. Springer, New York, NY, pp 77–113

Everest FA, Pohlmann KC (2001) The master handbook of acoustics, 5th edn. McGraw-Hill, New York, NY

Fletcher H, Munson WA (1933) Loudness, its definition, measurement and calculation. Bell Syst Tech J 12(4):377–430

Gardner WG, Martin KD (1994) HRTF measurements of KEMAR dummy-head microphone. MIT Media Lab Perceptual Computing – Technical Report #280

Geier M, Spors S (2012) Spatial Audio with the SoundScape Renderer. https://www.int.uni-rostock.de/fileadmin/user_upload/publications/spors/2012/Geier_TMT2012_SSR.pdf. Accessed 31 Dec 2014

Gelfand SA (2010) Hearing: an introduction to psychological and physiological acoustics, 5th edn. Informa, Essex

Gelfand SA, Levit H (1998) Hearing: an introduction to psychological and physiological acoustics. Marcel Dekker, New York, NY

Hafter ER, Sarampalis A, Loui P (2007) Auditory attention and filters. In: Yost WA, Popper AN, Fay RR (eds) Auditory perception of sound sources, vol 29, Springer handbook of auditory research. Springer, New York, NY, pp 115–142

Haneda Y, Makino S, Kaneda Y, Kitawaki N (1999) Common-acoustical-pole and zero modeling of head-related transfer functions. IEEE Trans Speech Audio Process 7(2):188–196

Hirst W, Kalmar D (1987) Characterizing attentional resources. J Exp Psychol Gen 116(1):68–81

Institut fur Nachrichtentechnik at Universitat Rostock (2014). http://www.int.uni-rostock.de/. Accessed 31 Dec 2014

Iqbal ST, Ju YC, Horvitz E (2010) Cars, calls, and cognition: investigating driving and divided attention. In: Proceedings of the 28th international conference on human factors in computing systems (CHI'10), Atlanta, GA, pp 1281–1290

Jacobson H (1951) The informational capacity of the human eye. Science 113(2933):292–293

Jo SD, Chun CJ, Kim HK, Jang SJ, Lee SP (2011) Crosstalk Cancellation for Spatial Sound Reproduction in Portable Devices with Stereo Loudspeakers, Communication and Networking. Springer, Berlin, pp 114–123

Kinsler LE, Frey AR, Coppens AB, Sanders JV (1999) Fundamentals of acoustics, 4th edn. Wiley, Hoboken, NJ

Kistler DJ, Wightman FL (1992) A model of head-related transfer functions based on principal components analysis and minimum-phase reconstruction. J Acoust Soc Am 91(3):1637–1647

Koch K, McLean J, Segev R, Freed MA, Berry MJ II, Balasubramanian V, Sterling P (2006) How much the eye tells the brain. Curr Biol 16(14):1428–1434

Kulkarni A, Colburn HS (2004) Infinite-impulse-response models of the headrelated transfer function. J Acoust Soc Am 115(4):1714–1728

Langendijk EHA, Bronkhorst AW (2002) Contribution of spectral cues to human sound localization. J Acoust Soc Am 112(4):1583–1595

Lee JD, Caven B, Haake S, Brown TL (2001) Speech-based interaction with in-vehicle computers: the effect of speech-based e-mail on drivers' attention to the roadway. Hum Factors: J Hum Factors Ergon Soc 43(4):631–640

Loki Software (2000) Loki Software, Inc. http://www.lokigames.com/. Accessed 31 Dec 2014

Macpherson EA, Middlebrooks JC (2002) Listener weighting of cues for lateral angle: the duplex theory of sound localization revisited. J Acoust Soc Am 111(5):2219–2236

Macpherson EA, Middlebrooks JC (2003) Vertical-plane sound localization probed with ripple-spectrum noise. J Acoust Soc Am 114(1):430–445

McKnight AJ, McKnight AS (1993) The effect of cellular phone use upon driver attention. Accid Anal Prev 25(3):259–265

Moller H, Sorensen MF, Hammershoi D, Jensen CB (1995) Head-related transfer functions of human subjects. J Audio Eng Soc 43(5):300–321

Moore BCJ (2004) An Introduction to the psychology of hearing, 5th edn. Elsevier, London

Miller JD, Wenzel EM (2002) Recent developments in SLAB: a software-based system for interactive spatial sound synthesis. In: Proceedings of the 2002 international conference on auditory display, pp 403–408

Navon D, Gopher D (1979) On the economy of the human processing system. Psychol Rev 86(3):214–255

Nees MA, Walker BN (2009) Auditory interfaces and sonification. In: Stephanidis C (ed) The Universal Access Handbook. Lawrence Erlbaum Associates, New York, NY, pp 507–522

OpenAL 1.1 (2005) OpeanAL 1.1 Specification and Reference. http://www.openal.org/documentation/openal-1.1-specification.pdf. Accessed 31 Dec 2014

OpenAL (2014) OpenAL from Wikipedia. http://en.wikipedia.org/wiki/OpenAL. Accessed 31 Dec 2014

OpenALSoft (2014) Sofwtare implementation of the OpenAL 3D audio API. http://www.openal-soft.org/. Accessed 31 Dec 2014

OpenGL (2014) OpenGL from Wikipedia. http://en.wikipedia.org/wiki/OpenGL. Accessed 31 Dec 2014

Oppenheim AV, Schafer RW, Buck JR (1989) Discrete-time signal processing, vol 2. Prentice-Hall, Englewood Cliffs, NJ

Performance Audio (2014) Performance Audio, Llc. http://www.performanceaudio.com/item/neumann-ku-100-dummy-head-microphone/2522/. Accessed 31 Dec 2014

Perrott DR, Sadralodabai TR, Saberi K, Strybel TZ (1991) Aurally aided visual search in the central visual field: effects of visual load and visual enhancement of the target. Hum Factors: J Hum Factors Ergon Soc 33:389–400

Quality and Usability Lab (2014) Institut für Softwaretechnik und Theoretische Informatik. http://www.qu.tu-berlin.de/. Accessed 31 Dec 2014

Ranney TA, Mazzae E, Garrott R, Goodman MJ (2000) Nhtsa driver distraction research: past, present, and future. In: Driver distraction internet forum, vol 2000

Rumsey F (2001) Spatial audio. Taylor & Francis, New York, NY

Slab (2014). http://slab3d.sonisphere.com/. Accessed 31 Dec 2014

Sodnik J (2007) The use of spatial sound in human-machine interaction. Dissertation, University of Ljubljana, Slovenia

Sodnik J, Susnik R, Stular M, Tomazic S (2005) Spatial sound resolution of an interpolated HRIR library. Appl Acoust 66(11):1219–1234

Sodnik J, Susnik R, Tomazic S (2006) Principal components of non-individualized head related transfer functions significant for azimuth perception. Acta Acustica united with Acustica 92(1):312–319

SoundScape Renderer (2014) Spatialaudio.net. http://spatialaudio.net/ssr/. Accessed 31 Dec 2014

Stifelman LJ (1994) The cocktail party effect in auditory interfaces: a study of simultaneous presentation. MIT Media Laboratory Technical Report, pp 1–18

Strutt JW (1907) On our perception of sound direction. Philos Mag 13:214–232

Susnik R, Sodnik J, Umek A, Tomazic S (2003) Spatial sound generation using HRTF created by the use of recursive filters. In: The IEEE Region 8 EUROCON 2003: computer as a tool, pp 449–453

Vliegen J, Van Opstal AJ (2004) The influence of duration and level on human sound localization. J Acoust Soc Am 115(4):1706–1713

White G, Louie GJ (2005) The audio dictionary: revised and expanded. University of Washington Press, Seattle, WA

Wickens CD (1984) Processing resources in attention. In: Parasuraman R, Davies R (eds) Varieties of attention. Academy Press, New York, NY, pp 63–102

Wightman FL, Kistler DJ (1997) Monaural sound localization revisited. J Acoust Soc Am 101(2):1050–1063

Young KL, Regan MA, Hammer M (2003) Driver distraction: a review of the literature. Technical Report 206, Monash University Accident Research Centre, Victoria, Australia

Zahorik P (2002a) Assessing auditory distance perception using virtual acoustics. J Acoust Soc Am 111(4):1832–1846

Zahorik P (2002b) Auditory display of sound source distance. In: Proceedings of the 2002 international conference on auditory display, Kyoto, Japan

Chapter 3
Auditory Interfaces

People are very social beings and interact with their environment in many ways (O'Shaughnessy 2003). Information can be received through seeing, hearing, smelling, tasting or feeling, while, conversely, it can be sent back into the environment visually, vocally, and through body gestures. Interaction between a user and a machine is performed through a user interface that consists of various hardware and software components. Its main functionality is to transform the users' orders and requirements into a set of machine commands and also to provide the corresponding feedback to the user, preferably in real-time. The available communication channels between a user and a machine are limited by human senses and their capabilities on the one hand and the technological restraints of input–output devices on the other. The primary communication channels for the majority of people are seeing and hearing, and consequentially visual and auditory interfaces are the dominant types of user interfaces in the HMI domain. The major paradigm of our times is undoubtedly mobility. The mobile life, such as driving a car, cycling or walking, requires a lot of visual attention. Interacting through a visual interface in these situations leads to distraction from one's primary task and consequently to a difficult, dangerous, ineffective and potentially frustrating user experience. In this chapter we present the basic properties and advantages of auditory interfaces that offer an excellent alternative to Human-Machine Interaction (HMI) in visually-demanding mobile situations. Auditory interfaces are very flexible and they do not interfere significantly with the reception and processing of visual information.

3.1 User Interfaces

In the visual domain, Graphical User Interfaces (GUIs prevail and allow users to interact with computers through graphical icons and more complex structures (Martinez 2011)). They successfully succeeded Command-Line Interfaces (CLIs) that required the user to know all the commands by heart and to enter them through a keyboard.

© The Author(s) 2015
J. Sodnik, S. Tomažič, *Spatial Auditory Human-Computer Interfaces*,
SpringerBriefs in Computer Science, DOI 10.1007/978-3-319-22111-3_3

Nowadays, GUIs represent a major component of all modern operating systems. GUIs can be displayed in a number of visual displays and screens, including head-mounted displays and head-up displays in vehicles. They are typically used through direct manipulation of various graphical elements and also often referred to as WIMP (Windows Icons Menus Pointers) interfaces. Different input devices can be used to interact with GUI, such as the computer keyboard (as well as the virtual keyboard used in modern mobile devices) and pointing devices (mouse, touchpad, joystick, etc.). GUI design approaches and interaction metaphors are also commonly adopted in various auditory displays, which will be described and evaluated in this chapter.

The sense of touch is also used in HMI, as it represents a base communication channel for so-called haptic or tactile displays (Answers 2014). By tactile we refer to touch-based interfaces where the texture of a surface or selected device holds certain information about the system. Braille keyboards, for example, belong to the group of tactile devices that provide a tactile reading system for visually-impaired and blind users. Haptic interfaces, on the other hand, are force-based and require human muscles to be involved for the perception of information. Virtual reality systems often include haptic feedback devices to increase the realism of perception or, for example, force-feedback joysticks and steering wheels in gaming and simulation software.

The third major group is represented by auditory interfaces which use sound and the human auditory channel to exchange information between users and various types of computers and information systems. They are bidirectional interfaces that provide input and output mechanisms. For the input, machine listening and speech recognition systems are used to capture sound events and transform them to computer commands. Output technologies are often referred to as auditory displays that transmit information from a system to a user.

3.1.1   General Properties of Auditory Interfaces

Auditory interfaces have gained importance due to the growing need for ubiquitous or embedded interactivity. These terms refer to interactivity with mobile and wearable devices with a wide set of functionalities and usage scenarios where visual interaction is difficult and display size is limited. Mobile devices tend to be smaller and lighter, forcing the display sizes to be kept to a minimum or even abandoned. Another common problem of ubiquitous interactivity is the user's occupation with another primary activity while the interaction with intelligent devices is of secondary importance. Typical examples of such primary activities are walking, running, cycling, or driving, where human visual attention is focused on navigation and orientation. Visual interaction, such as reading from a screen and processing the content, is significantly impaired while also causing a significant distraction to the primary activity. Auditory interfaces require much less physical activity and rely mostly on the hardware, with processing and memory capacity coupled with the microphone or speaker(s). Processing and memory units are becoming smaller and cheaper and are do not limit the final size of the mobile device.

Auditory displays have been used for many decades, mainly as alarms and feedback tools (Peres et al. 2008). The Morse code used in early communications is an example that dates back to the 1800s. The majority of modern auditory interfaces are still limited to very simple warning sounds and beeps that inform the user about some background processes or events which require user action (Gardenfors 2001). Examples of such events are opening or closing a program, pressing keyboard or mouse buttons, incoming emails, operating system errors, low battery levels, etc. In the case of mobile phones and other mobile devices, sound alarms are vital for attracting immediate attention, for example in the case of received messages, emails, or phone calls. Some modern operating systems use additional, albeit optional, sounds and music clips to provide a multimedia experience and to enrich overall user satisfaction. In some cases sounds can be used to communicate ongoing processes such as downloading or uploading files, deleting files, emptying the recycle bin, etc. Computer games and multimedia software use more advanced sounds, including multichannel and spatial audio. In these cases sound is used to enhance the user's immersion and to improve interaction.

Sounds with crucial information about some ongoing activities have different attributes than sounds used just as complement visual information (Gardenfors 2001). The first group of sounds represents the main feedback about the system (such as auditory interfaces for visually impaired people), in-vehicle information systems, telephone-based interfaces, portable hands-free devices, etc. These sounds are played frequently and sometimes very loudly, and should therefore be very comprehensive and simple in order not to be noisy or annoying.

The second group of sounds is used mainly to complement visual interfaces and to enhance the HMI by making it more natural and pleasant. Due to different properties and limitations of human visual and auditory channels, each interface group addresses different aspects of interaction. The visual channel enables efficient detection of specific and detailed information within a small area of focus. Sounds on the other hand can represent subtle and non-disturbing background events. Music in movies can, for example, convey information to the audience without interference with the main video content. Sometimes the sound can be so subtle it is only perceived subconsciously.

Auditory interfaces as they are experienced now can be more advanced and sophisticated primarily due to the increasing availability of processing power and memory, which provides the technical background for creating these interfaces (Peres et al. 2008). This field of research is currently experiencing exponential growth, but still lacks the lexicon and taxonomy for different types of auditory interfaces.

There are three different classifications of auditory displays, divided based on the application, user, and sound orientation (Johannsen 2004). Application-oriented classification divides auditory interfaces into:

- vehicular guidance: aircraft and automotive,
- medicine: intensive care units, operation theaters, monitoring of life supporting functions,
- cinemas and the entertainment industry,

- industrial plants,
- mobile devices,
- etc.

User-oriented classification divides auditory interfaces into several dimensions concerning different classes of users, communication with individual preferences and needs, and individual users' capabilities (Johannsen 2004). Users can, for example, be classified as:

- operators,
- engineers,
- managers,
- sportsmen,
- etc.

Their individual capabilities can, for example, additionally classify them as:

- experts,
- novice users,
- occasional users,
- disabled users,
- musicians or non-musicians,
- etc.

The third classification is based on sound orientation and divides auditory interfaces into speech-based and non-speech-based interfaces. Each group provides different functionalities and offers different advantages and disadvantages. In this chapter we use the sound-oriented classification and summarize the basic properties of speech-based and non-speech-based auditory interfaces. In Chap. 4, where we focus particularly on spatial auditory interfaces, we use the application-oriented classification.

3.2 Speech Interfaces

Speech interfaces are based on human speech which can be detected and recognized, recorded and replayed or synthesized by the computer. Speech is the fundamental and most natural mode of communication between humans and is therefore learned from an early age. The main idea and goal of speech interfaces is to provide this natural communication and interaction with computers, mobile devices, wearable devices, home appliance, etc., as well. Ideally the user is not required to learn how to use an individual interface or to memorize any commands or interaction patterns. On the other hand, if the language supported by the interface is not understood by the user, the interface is more or less useless.

There are at least two fundamental advantages of speech (Rosenfeld et al. 2001). It can be described as an ambient medium that allows simultaneous speech and visual interaction without severe interference. Speech has also proved to be more descriptive than referential, requiring objects of interaction to be described by their properties. In visual interaction the same objects can simply be just pointed to or grasped. It is therefore possible to complement and combine speech and pointing as multimodal interaction.

Simple and less-intelligent machines and devices are the most convenient for use with speech interfaces (Rosenfeld et al. 2001). In this case, a user can get a mental model of a machine's capabilities and its current state through a vocal description. Unfortunately, a user often has to adapt to the "machine's natural language" and produce a set of simple and predefined voice commands and instructions.

3.2.1 Technical Background

The two mechanisms related to HMI via speech are Text-To-Speech (TTS) synthesis and Automatic Speech Recognition (ASR). We summarize some basic properties of the two processes based on O'Shaughnessy (2003). TTS simulates a person speaking, while ASR simulates a person who is listening. They represent a technological ground for transforming speech to text and text to speech, respectively. Understanding speech and its content is also vital for successful TTS and ASR.

ASR has proved to be more difficult than TTS due to the unreliable segmentation of recorded speech into smaller units, such as words and phrases, or in some cases phonemes. These small linguistic units have complex relationships to aspects of the acoustic speech signal. TTS, on the other hand, receives text as an input and divides it to letters and words. The segmentation of text is in general much simpler than segmentation of live speech. Another issue contributing to the different difficulties of these two processes is the human capability to adapt. Having a human listener as a receiver and interpreter of speech is a great advantage due to the simple and almost subconscious adaptation to an unusual accent or poor quality of synthesized speech. In the case of ASR, machine algorithms face many difficulties adapting to various speakers, their accents, individual vocal properties, talking pace, etc. People never say the same thing in exactly the same way twice.

Speech recognition and synthesis mimic human behavior and require a design approach based on Artificial Intelligence (AI). For the AI it is important to understand the mechanisms used by humans to perform and accomplish individual intelligent tasks. Unfortunately, the majority of intelligent tasks are performed by the brain, which is still the least understood organ—its internal processes related to different activities are mostly unexplained. Therefore, AI research focuses predominantly on the functioning of peripheral organs and their role in the processing activities of intelligent tasks. In the case of ASR and TTS processes, it is therefore well know what is happening in different regions of the human ear and also different regions of the vocal organ. The role of the brain in converting linguistic concepts into the muscle activity of the vocal organ or in interpreting electrical signals from a large number of hair cells in the inner ears is almost entirely unknown.

An important building block of all ARS and TTS systems is their vocabulary. The majority of languages consist of thousands of words, forming trillions of different sentences. It is therefore impossible to record and store all possible sentences. In practice, a typical vocabulary for speech synthesis or recognition stores units smaller than full sentences, for example words or even smaller linguistic units. A set of words is limited by the dictionary defined for an individual speech application, which also has to store their pronunciations. It is always limited and should be known and obeyed by the user speech interface. However it is always difficult to avoid the appearance of unknown words and phrases produced by users. In the case of TTS, there are different universal synthesis algorithms that handle words which are not contained in the dictionary, while in the case of ASR the unknown speech is labeled as out-of-vocabulary and ignored. In the future the size of dictionaries is expected to increase due to larger capacities of memory units.

Speech communication can also sometimes take place when audible acoustic signals are unavailable (Denby et al. 2010). Such systems, called Silent Speech Interfaces (SSI), could be used as an aid for the speech handicapped or as parts of communication systems in silent or very noisy environments. SSIs are currently still in the experimental stage based on several different types of technology. The most promising technological approaches include the capture of movement of fixed points on the articulators using electromagnetic articulography, performing real-time characterization of the vocal tract with ultrasound, optical imaging of tongue and lips, etc. The details on these technologies and several others proposed for SSI can be found in Denby et al. (2010), although they all require significant advances in instrumentation and signal processing in order to become fully functional.

The technical background and details of individual processing steps in speech interfaces can be found in O'Shaughnessy (2003). Nowadays commercial speech synthesizers are widely available for all major world languages with very high speech quality and intelligibility. The development of processing power and increased memory capacity have made it possible for synthesizers to run on any software.

3.2.2 Design Alternatives

There are several possible approaches to the creation of human-friendly speech systems (Rosenfeld et al. 2001). The ideal case would be to allow the user to use unconstrained natural language with spontaneously-formulated speech for interaction with a device, while at the same time knowing only the domain of operation and its specifics. Since in this case the user is not forced to adapt to any limitation of the speech interface, it is up to designers and developers of the interface to deal with the unlimited set of spoken words and phrases. The task is nearly impossible, as all computing systems and humans have limited knowledge, processing and memory capacity.

A more realistic approach is to limit the user with a predefined dictionary of available commands and words. Two examples of such limited speech interfaces are dialog trees and command and control interfaces. In the case of dialog trees, domain activities are broken down into a sequence of hierarchically-organized levels with several choices. At each level a user can select only one command from several available options, which should always be presented to a user when the individual level within the tree is reached. The main drawback of such systems is their tree structure, which prevents a user from skipping several levels and directly accessing the point of interest. It is also difficult for the designers to add new functionality or new choices to the system, since it requires the rebuilding of the entire tree structure.

Command and control interfaces also reduce complexity by only defining a short dictionary of possible phrases. The set of phrases is then combined with custom input variables to form various speech commands. One example of such an interface is an in-vehicle speech interface for operating a mobile phone where a driver can simply say "CALL" and add the name of the selected person from the contacts list. Such interaction can be very efficient, presuming that the user has the necessary knowledge about the domain in which the system operates. However, command and control interfaces require the user to learn the language of the system, which makes them unfeasible for multiple applications that are different and operating domains.

3.2.3   Constraints and Limitations

In practice, there are several important limitations and issues related to modern speech interfaces. The input interfaces are based on voice- or speech-recognition mechanisms which translate individual words or voice commands into a set of computer instructions and input commands. The main problem of speech recognition systems is their relatively low robustness and accuracy. It is nearly impossible to build a robust recognition system which can be used in heterogonous environments with unpredicted and highly-dynamic noise conditions. The robustness also decreases as the number of different users and voices is increased, due to individual vocal properties and limitations. The majority of speech recognition systems work very well in a controlled and limited environment, with a limited number of users and limited number of available and recognizable voice commands and phrases.

The lack of robustness and the low recognition ratio can be frustrating and unpleasant for users, offering a bad user experience and general dissatisfaction with the entire system. The other major problem of speech interfaces is a lack of privacy and anonymity for users. Speech commands and phrases have to be spoken aloud and very clearly in order to be recognized correctly. They are therefore very unsuitable for offices and other rooms with multiple users.

On the other hand speech is a slow mode of communication (Gardenfors 2001). It is very inconvenient to use speech to accompany ongoing processes. No one would, for example, be interested in listening repeatedly to the words "copying, copying,

copying" while a file is being copied from one location to another. Speech should therefore be avoided as routine task feedback. It is also very precise and requires high levels of human attention and focus to be processed and understood successfully. Spoken sentences have to be heard from the beginning to the end in order for their meaning and message to be perceived correctly. Sometimes many words have to be used to describe something very simple and with low information content.

Due to above-mentioned constrains speech interfaces have still not reached their full potential (Rosenfeld et al. 2001). Their impact has currently been limited to places where their utilization offers great advantages over existing user interfaces. The main limitations which should be addressed and improved in the future are recognition performance, accessible language for all users, and a complicated and sophisticated development process.

3.3 Non-speech Interfaces

Non-speech auditory interfaces use natural or artificially-created sounds to describe different types of information, symbols or metaphors. There are several different classifications or categorizations of non-speech interfaces based on the choice of the used sounds, the method of mapping sound to information, or the functionality that the sounds have provided (Gaver 1997). The choice of sounds used in auditory interfaces includes simple multidimensional tones, music tones, and everyday natural sounds. The way sound is mapped to the information can be completely arbitrary and non-intuitive, or very metaphorical and intuitive. The functionalities of non-speech interfaces vary from simple alarms and warning systems to the exploration of complex multidimensional data. Sounds, mapping, and functions are closely related and form different areas of the space.

3.3.1 Sonification

A typical form of using non-speech sounds for representing complex and multidimensional data is data auralization or sonification (Gaver 1997; Peres et al. 2008). Sounds can be used to simply enhance visual display or to represent an independent interface by mapping data parameters to sound parameters. These can be individual tones or a continuous stream manipulated by the changes in the data. Simple sonification is often used in computer programs and operating systems or various monitoring systems that require the user's attention at critical moments.

Multidimensional data that consists of a large number of points should be sonified as an integral whole, allowing a user to perceive various high-level patterns. Let us imagine an example of the sonification of temperature and humidity levels in different rooms in a building. Key issues related to such complex mapping are:

- how to choose the most appropriate set of sounds to be associated with the variable they present (e.g., should each room be represented with different characteristic music?),
- how should changes in variables be reflected in sounds (e.g., by changing pitch, amplitude, position),
- how sound properties should be changed (e.g., increased or decreased),
- what the response time and scale should be (e.g., how much should sound pitch change if the temperature drops by 1 °C?).

In graphical interfaces, a typical example of unified representation of multidimensional data is a 2D or 3D scatter plot. Each point on the curve represents two or three dimensions simultaneously. Additional information can be added through different properties, such as color, size, the shape of the curves, etc. A similar approach could also be used when describing such complex data with multidimensional sound, but no such tools have been developed or proposed so far. The main problem with sound is its strong serial nature. While the visual sense can process data in parallel and always focus on one particular bit of information, a human auditory system is only able to perceive and process a very limited number of simultaneous sound streams. In an attempt to present scatter plots aurally, various reference tones could, for example, indicate values in individual dimensions. If there were many different dimensions it would be impossible to play all tones simultaneously. Additionally, playing individual tones sequentially is also problematic due to auditory memory limitations.

3.3.2 Auditory Icons, Earcons and Spearcons

In some cases auditory interfaces can also replace GUIs and more semantic interfaces which use icons to represent different applications or their in-built features. Gaver (1986) proposed the use of natural everyday sounds to represent various components of a computer interface. The auditory equivalents of visual icons are called auditory cues, which translate different symbols into auditory artifacts. Based on their abstraction level, auditory cues belong to one of the three groups described below.

The cues of the first iconic group are called auditory icons, which try to represent an event as realistically as possible. An example of an auditory icon is the sound of water being poured from one glass to another to represent a file being copied from one location to another or downloaded from the internet.

The second group, called earcons, represents icons or functions with more abstract and symbolic sounds. The relation between an event and a corresponding sound is completely arbitrary and does not allow any semantic relation. The meaning of earcons has to be learned for each individual interface. Examples of earcons are typical computer or mobile phone sounds indicating various background events and notifications (e.g., the arrival of a new email, shutting down a system, changing the active profile on a mobile device, etc.). Earcons can be designed to represent not only a single item, but also its position in a more complex hierarchical structure (Brewster et al. 1993) in different types of auditory or multi-modal interfaces. Comparative studies of auditory icons and earcons showed no significant difference in efficiency between the two (Brewster 2001; Lucas 1994).

The third group of auditory cues is called spearcons, created by speeding up spoken texts and phrases until they are no longer recognized as speech (Walker et al. 2006). The speed of the reply can be slowed down at the beginning to facilitate learning and then sped up after associations between auditory cues and interface commands are made. Spearcons have proved to be very efficient for sonification of hierarchical menu structures, where they represent individual menu items at different levels.

All types of auditory cues can be efficiently used to represent a hierarchical menu structure and the corresponding events. The amount of information represented by a single sound event is always limited due to limited capabilities of the human auditory system. One way of expanding these limitations and increasing information flow is by using spatial sound or the spatial distribution of multiple sound sources. Auditory interfaces utilizing spatial sound are presented in the next chapter. They use standard auditory interface design principles and methods and extend them by using spatial sound originating from different positions in space.

3.3.3   Interface Metaphors

According to Wikipedia (Metaphor 2014) "a metaphor is a figure of speech that describes a subject by asserting that it is, on some point of comparison, the same as another otherwise unrelated object". In user interfaces, metaphors are used to give a user immediate knowledge about the meaning of individual interface elements and about how to interact with the system (Reimer 2005). They are based on a set of visual or aural representations of interface tasks and procedures known to a user from other domains. They can define the representation and outlook of the interface or the corresponding interaction mechanism. Metaphors can also be an effective mechanism for developing and supporting the user's mental model of the interface. An example of a visual interface metaphor is an icon of a trash bin representing a folder with deleted content. It is a part of a desktop metaphor which is well established in the majority of modern operating systems. In this structural metaphor, files and folders are also represented with icons and can be dragged and dropped in different locations. Dropping a file icon onto the trash bin icon represents the action of deleting the selected file.

There are no such strong and intuitive interface metaphors for auditory interfaces, however many different ideas have been proposed and evaluated. A ring and dial metaphor has, for example, been proposed to create a browsing environment for audio recordings (Kobayashi and Schmandt 1997), notifying a user about important events such as incoming emails, voicemails, calendar entries (Sawhney and Schmandt 2000), for navigating between different sound sources (Crispien et al.

1996), or for navigating a hierarchical menu structure (Sodnik et al. 2008, 2011). In this type of metaphor a user is usually positioned in the center of a ring while different sound elements and streams orbit around the user's head. The user can select individual elements through different interaction mechanisms. An interesting and also very extensive auditory interface based on several ontological metaphors was Audio Windows (Cohen and Ludwig 1991). The entire system was represented as a room and individual objects and icons are represented as sound cues. They could be manipulated through different interaction techniques, including a data glove as the input device. A similar approach was also used in (Gaver et al. 1991), where a system was represented as an office or as a multiple room metaphor (Aoki et al. 2003). Most of the proposed above-mentioned auditory interfaces incorporate spatial sound, and therefore some are explained in more detail in Chap. 4.

3.4 Conclusion

In the past, auditory interfaces have mainly been used to complement visual interfaces as alarms and feedback tools that acquire the user's attention at a given time. Nowadays, the mobility in everyday life requires devices to be smaller and lighter in order to be carried around and used anywhere and anytime. The ever-smaller displays and keyboards fundamentally change the way we think about user interfaces for these devices and HMI in general. Auditory interfaces offer an excellent alternative for exchanging information between a device and a mobile user or a user engaged in another sort of primary activity. With the rapid development of technologies and tools for recognition and synthesis of human speech, speech-based auditory interfaces have matured enough to be successfully integrated in many commercially-available systems and applications. Auditory interfaces will probably never fully replace visual interfaces, but rather complement them. Future mobile devices will most likely not rely on one single interaction or display technique, but instead offer a variety of simultaneous options from which the users can choose, depending on the type of information they are accessing and on their current situation.

References

Aoki PM, Romaine M, Szymanski MH, Thornton JD, Wilson D, Woodruff A (2003) The mad hatter's cocktail party: a social mobile audio space supporting multiple simultaneous conversations. In: Proceedings of the SIGCHI conference on Human factors in computing systems (CHI'03), Ft. Lauderdale, FL, pp 425–432

Answers (2014) What is the actual difference between haptic and tactile? http://www.answers.com/Q/What_is_the_actual_difference_between_haptic_and_tactile. Accessed 31 Dec 2014

Brewster S (2001) Providing a structured method for integrating non-speech audio into human-computer interfaces. Dissertation, University of York, Great Britain

Brewster SA, Wright PC, Edwards AD (1993) An evaluation of earcons for use in auditory human-computer interfaces. In: Proceedings of the INTERACT'93 and CHI'93 conference on Human factors in computing systems, New York, NY, pp 222–227

Cohen M, Ludwig LF (1991) Multidimensional audio window management. Int J Man Mach Stud 34(3):319–336

Crispien K, Fellbaum K, Savidis A, Stephanidis C (1996) A 3D auditory environment for hierarchical navigation in non-visual interaction. In: Proceedings of the International Conference on Audio Display (ICAD'96), Palo Alto, CA, pp 18–21

Denby B, Schultz T, Honda K, Hueber T, Gilbert JM, Brumberg JS (2010) Silent speech interfaces. Speech Comm 52(4):270–287

Gardenfors D (2001) Auditory interfaces: a design platform. http://jld.se/dsounds/auditoryinterfaces.pdf. Accesed 31 Dec 2014

Gaver WW (1986) Auditory icons: using sound in computer interfaces. Hum Comput Interact 2(2):67–94

Gaver WW (1997) Auditory interfaces. Handb Hum Comput Interact 1:1003–1041

Gaver WW, Smith RB, O'Shea T (1991) Effective sounds in complex systems: the arkola simulation. In: Proceedings of the SIGCHI Conference on Human Factors in Computing Systems (CHI'91), New Orleans, LA, pp 85–90

Johannsen G (2004) Auditory displays in human-machine interfaces. Proc IEEE 92(4):742–758

Kobayashi M, Schmandt C (1997) Dynamic soundscape: mapping time to space for audio browsing. In: Extended abstracts on human factors in computing systems (CHI'97), Atlanta, GA, pp 194–201

Lucas P (1994) An evaluation of the communicative ability of auditory icons and earcons. In: Proceedings of the second international conference on auditory display, Santa Fe, pp 121–128

Martinez WL (2011) Graphical user interfaces. Wiley Interdiscip Rev Comput Stat 3(2):119–133

Metaphor (2014) Metaphor from Wikipedia. http://en.wikipedia.org/wiki/Metaphor. Accessed 31 Dec 2014

O'Shaughnessy D (2003) Interacting with computers by voice: automatic speech recognition and synthesis. Proc IEEE 91(9):1272–1305

Peres SC, Best V, Brock D, Frauenberger C, Hermann T, Neuhoff JG, Nickerson LV, Shinn-Cunningham B, Stockman A (2008) Auditory interfaces. HCI beyond the GUI: design for haptic, speech, olfactory, and other nontraditional interfaces, pp 147–195

Reimer J (2005) A History of the GUI. Arstechnica.com

Rosenfeld R, Olsen D, Rudnicky A (2001) Universal speech interfaces. Interactions 8(6):34–44

Sawhney N, Schmandt C (2000) Nomadic radio: speech and audio interaction for contextual messaging in nomadic environments. ACM Trans Comput Hum Interact 7(3):353–383

Sodnik J, Dicke C, Tomažič S, Billinghurst M (2008) A user study of auditory versus visual interfaces for use while driving. Int J Hum Comput Stud 66(5):318–332

Sodnik J, Jakus G, Tomažič S (2011) Multiple spatial sounds in hierarchical menu navigation for visually impaired computer users. Int J Hum Comput Stud 69(1):100–112

Walker BN, Nance A, Lindsay J (2006) Spearcons: speech-based earcons improve navigation performance in auditory menus. In: Proceedings of the international conference on auditory display (ICAD 2006), London, England, pp 63–68

Chapter 4
Spatial Auditory Interfaces

Spatial auditory interfaces use three-dimensional sound as an additional display dimension and consist of audio items at different spatial locations. They have evolved significantly in the last couple of years and can be found in a variety of environments where visual communication is obstructed or completely blocked by other activities, such as walking, driving, flying, operating multimodal virtual displays, etc. The precise spatial position of each source can offer an additional informational cue in the interface or can simply help resolving various ambiguities in the content of simultaneously-played sources. It can also be used to increase realism in virtual worlds by imitating real environments where the majority of sounds can be localized and associated with their sources.

Humans' localization principles and the important techniques for the generation and reproduction of spatial sound have been explained and discussed in Chap. 2. In this chapter we summarize the most important and often-cited research that reports on spatial auditory interfaces in different contexts and for different tasks. Separate chapters are dedicated to the most important areas of spatial auditory displays: portable devices and computers, virtual environments, aircrafts and vehicles, visually impaired and blind computers users and brain-computer interfaces.

4.1 Portable Devices and Music Players

The majority of user interfaces for desktop computers, laptops, and mobile devices are multimodal, combining visual and audio displays. Interaction with these systems is based mostly on visual displays and corresponding GUIs. Sound, on the other hand, only plays a minor role and serves mostly as a notification or a warning system or to give feedback about the user's input. Mobile devices have restricted input and output capabilities compared with desktop computers and their usability

© The Author(s) 2015
J. Sodnik, S. Tomažič, *Spatial Auditory Human-Computer Interfaces*,
SpringerBriefs in Computer Science, DOI 10.1007/978-3-319-22111-3_4

is reduced. They suffer from a very limited amount of screen space and the visual display can become cluttered with various notifications and widgets. Input is also limited due to small keyboards and buttons. New interaction techniques are therefore being researched and evaluated in order to effectively utilize new services while being mobile. An important segment of the portable device market belongs to portable music players and similar audio-playing devices. They all require a user to browse through a collection of available songs and audio clips to select the desired content. Several researchers investigated the possibility of using a spatial auditory interface as a feedback mechanism in such interactions with different touch-based input interfaces.

Schmandt (1995) proposed an auditory interface called AudioStreamer as one of the earliest attempts of taking advantage of the cocktail party effect and exploiting users' ability to separate the mixture of sounds arriving at the ears at a certain point in time. It consists of three simultaneous sound sources enriched by various acoustic cues (e.g., location, harmonics and frequency, volume, continuity, etc.) which allow their separation into distinct sources. The sources used are three different news anchors played from three different spatial locations. The listener's ability to browse between these sources is further enhanced by tracking the user's head and increasing the gain of the source currently faced by the user. Evaluation of the AudioStreamer revealed that when the user focuses on one of the available simultaneous channels, almost none of the information is perceived from the other two channels. Therefore, an additional feature was built into the AudioStreamer using various acoustic cues to alert user about salient events on one of the unselected channels (e.g., story boundary—a 400 Hz, 100 ms tone with the increased gain of 10 dB).

A similar project, inspired by the AudioStreamer, is called Dynamic Soundscape (Kobayashi and Schmandt 1997). It is a browsing environment for audio recordings and it is based on a sound source called "Speaker" orbiting the user's head and playing different sound streams. Each sound stream is fixed at a certain spatial angle and starts playing when "Speaker" reaches its location (see Fig. 4.1).

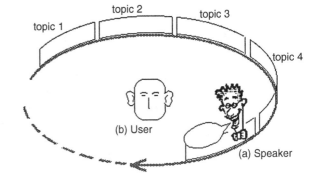

Fig. 4.1 Dynamic Soundscape: simultaneous presentation and spatial mapping of audio streams (Kobayashi and Schmandt 1997)

The user can replay the selected topic or jump ahead to the new topic by pointing in the direction of its source with the touchpad interface attached to the surface. The pointing gesture creates a second "Speaker" at the selected location, which starts playing the selected source while the original "Speaker" keeps orbiting and playing other sources with decreased loudness. There is often a mismatch between the location selected by the user and the correct location of the targeted sound source. The authors have proposed a solution in the form of an "audio cursor"—an additional sound source in the virtual space that provides constant feedback of the user's selected location. In this way, the user has to acoustically overlay the "audio cursor" and the targeted sound source in order to perform the selection. Evaluation of the proposed system revealed the significant importance and usability of spatial memory for audio navigation, as users can memorize individual topics by their spatial location.

Schmandt (1998) also proposed another spatial auditory interface for browsing collections of sound files called the Audio Hallway. The user browses the files by traveling up and down a hallway, passing rooms on the left and right. A specialized auditory collage of "braided audio" indicates the content of each room. Rooms are represented as broadcast radio news stories with a collection of individual "sound bites." When selecting and entering an individual room, the sounds of the room are arranged spatially in front of the user in the shape of an arc. The user controls the auditory focus with head rotation, where the sound files in the user's focus become louder. The user's navigation in the hallway is based on head movements, such as tilting forward, backward, left, or right. The author reported disappointment at how difficult it was to provide the desired auditory experience while traveling down the hallway due to the lack of visual information and cues.

Touch Player (Pirhonen et al. 2002), on the other hand, is a dedicated spatial auditory interface for controlling a portable music player—simulated by a pocket PC. Sounds are used primarily to give feedback on the preformed input gestures and consist of spatialized earcons. They are placed on a horizontal line in front of the user using a straightforward stereo panning technique. The most important novelty of this project is the proposed gesture-based input. Gestures are drawn on a special touch screen that supports the basic functions of the player: play/stop, next/previous song, volume up/down, etc. Swiping across the screen from left to right or from right to left changes the track, a single tap starts or stops the player and an up/down swipe changes the volume. The authors proved that it is possible to create a tactile/auditory interface for a mobile music player which can be used without a visual interface. The explicit and immediate audio feedback on the gestures is very important to making the user comfortable and confident in its use.

As a continuation of this work, Brewster et al. (2003) proposed an extension of Touch Player which consisted of a wearable PDA attached to the user's belt, a pair of headphones, and a head tracker. It also uses 2D gestures drawn on a screen as input and audio as output. The sound is also spatialized in a plane surrounding the user's head at the level of the ears. The head tracker is used to detect head orientation and to respatialize the sounds accordingly. It also enables the detection of head movements such as nods and shakes, which can be used as an additional interaction technique. Each audio source is located at its own angle relative to the user and can be selected by nodding toward an earcon and moving it to the front-center of the user. The non-important items can be moved to the rear by pointing at them with the PDA or head. Several simultaneous sounds can be played at different positions, such as a sonification of some processes on one side and a live synthesized speech stream on the other.

The Apple's iPod is undoubtedly one of the most popular portable audio players on the marker with its unique circular touch input interface. Shengdong et al. (2007) proposed the EarPod, a portable audio device using a very similar touch input interface, but which also provides synchronized audio feedback. The touchpad is divided into outer and inner rings, each of which is further divided into sectors. By touching the individual sector, the user can hear the sound of the corresponding menu item, while by continuously sliding a finger around the circle he or she can browse all available menu items. The user can select a menu item by simply lifting their finger at the desired sector. All speech-based audio feedback is spatialized in order to additionally map the menu items to the circular touch-sensitive surface. Due to the limitations of processing power on mobile devices, a simple left–right spatialization is used. It proved to be accurate enough to create the illusion that items were laid out in a circle around the user's head. The most important feature of the proposed system is its simple hardware, which only requires a circular touchpad and stereo headphones. Therefore, it can easily be deployed with existing devices with touchpads (e.g., iPods, cell phones, etc.).

As the majority of the above-mentioned interfaces used a ring metaphor for presenting auditory cues to a user, Hiipakka and Gaëtan (2003) proposed a slightly different spatial auditory interface for interaction with an aurally-presented structured collection of music files. The collection is organized as a hierarchical menu which includes several levels, e.g., music style, artist, album, and song. The process of selecting songs in a music collection consists of navigation, search, and selection. The proposed interface emulates the presentation of a 2D menu structure and uses vertical representation of the menu and corresponding items. Three items within each level of the menu are available simultaneously. A user should be able to select any item from any level of the hierarchy. The navigation is based on arrow keys, using up/down keys to change the level and left/right keys to select the item within the level. A prototype implementation of the interface on a PDA demonstrated its efficiency and showed great potential for future work.

Nowadays, a music player is often just one of the many available features on smart phones or other smart mobile devices. They are exposed to significant increases in processing power and network bandwidth and are expected to provide a variety of new multimedia services. They also require new interaction mechanisms and interfaces to support a variety of complex input and output commands.

Nomadic Radio (Sawhney and Schmandt 2000) is an interesting example of a complex wearable messaging application for managing voice and text-based messages, emails, calendar entries, news, etc. Messages are dynamically structured within different categories enabling a user to select a category, browse messages sequentially or save them and delete them on the server. Nomadic Radio uses spatial audio for positioning sound sources around the user's head that represent different messages and text entries; the sequence of the positions depends on the chronological order of messages' arrival in the system. The system is operated by speech commands and optional tactile input. The latter is used primarily to enable or disable the speech recognition system (e.g., push-to-talk mode) in noisy environments and to increase its robustness and recognition rate. It is based on a shoulder-mounted pair of directional speakers and a directional microphone placed on the chest. The volume of the speakers is adjustable to ensure that they are heard primarily by the user and not by the environment. This prototype of the wearable audio system is connected to a portable PC which communicates with the backend server via a wireless network.

Diary in the Sky (Walker et al. 2001), on the other hand, is a prototype of an auditory version of a calendar on a PDA. On the visual display the individual events in the calendar are displayed in 1-hour denominations in a vertical list. The proposed auditory version of the display presents calendar entries as spatial sound sources positioned on a clock-like layout surrounding the user's head in the horizontal plain. 12 am/pm is positioned directly in front of the user while 9 am/pm and 3 am/pm are positioned on the left- and right-hand side, respectively. Participants of the evaluation experiment appreciated the proposed solution and found it pleasant and easy to use. The authors also intended to explore the technique of using spatial sound by encoding message semantics into the spatialized audio streams. This is a general display technique which could be used in different interfaces and applications.

Another important set of features of portable smart devices is navigation- and location-based services. They also require customized interfaces that support the exchange of this specific type of information. AudioGPS (Holland et al. 2002) is an example of spatial auditory interface for non-visual interaction with the Global Positioning System (GPS). It allows users to be engaged in a specific location-based task while their eyes are unavailable due to occupation with another primary activity or task. The GPS system is used for navigation output for two important types of information: the direction in which the destination lies relative to the current direction of motion, and the distance to the destination. A briefly-repeated tone is used as a sound source, which is then panned left or right to indicate the direction of the target. The panning effect enables a simple distinction between left and right directions, while front and back directions cannot be perceived. Therefore, two versions of tone, i.e., sharp and muffled, are used to indicate the front and back direction, respectively. The direction the user is currently facing is represented by the direction of travel. Distance, on the other hand, is encoded as a "Geiger counter" that indicates the distance by the frequency of repetitive tones (e.g., a higher frequency means a closer distance). An additional arrival tone is played when users reaches the target destination.

Kan et al. (2004) also report the use of spatial audio for representing locations of users in a mobile communication system. The system consists of two laptops with GPS modules, microphones, and pairs of stereo headphones which enable two users to communicate by hearing each other from the corresponding spatial direction. GPS modules output information on their own positions and their waypoints. The waypoint of each module is another's module location. The user's head orientation is calculated from the direction of motion by taking into account the last 20 recorded positions. Based on both the users' head orientations and the bearing angles (e.g., the angle to the waypoint), the correct relative direction can be calculated and the sound stream (i.e., the speech signal of another user) can be played from the corresponding spatial angle.

Several researchers have proposed some general approaches to the creation of spatial auditory interfaces for various devices, platforms, and very specific tasks. Walker and Brewster (2000) proposed a spatialized audio progress bar for monitoring a number of background tasks and processes which only used sound to report on the rate, progress, and also to indicate completion or interruption. The goal is to provide these cues to users without interrupting them and without drawing their attention away from a foreground task. The progress bar consists of two major components: a fixed reference axis and a moving progress indicator. These components are presented with two non-speech spatial sounds. The reference sound is fixed in front of the user while the "progress" sound changes its spatial position on a circular orbit and communicates task completion by its angular location. The rate of progress is indicated simply by the angular speed of the orbiting sound. Two identical sounds played from in front of the user indicate the endpoint. The authors wanted to prove that spatial audio can be a key ingredient in multitasking interface design by enabling perception of multiple simultaneous sound streams.

Streaming Speech (Goose et al. 2002), on the other hand, is a framework for creating multiple simultaneous audio objects which are then spatialized in 3D space, multiplexed into a single stereo audio signal and streamed to a mobile device. The framework is based on the proposed 3D audio extensions to the Synchronized Multimedia Integration Language (SMIL)—an XML recommended by the World Wide Web Consortium for describing multimedia presentations (Smil 2014). Streaming Speech also consists of a server-based framework which can receive a SMIL file as an input and dynamically create a 3D audio stream as output. It represents a perfect solution for delivering and rendering rich 3D audio content—described by a markup language—to any commercially-available mobile device via a wireless network.

We have already described the process of sonification as an important aspect of auditory interfaces. Spatial sound and spatial auditory interfaces can also be used for representation and sonification of spatial and non-spatial data. The goal is to give the user nominal, qualitative, or quantitative judgment of the perceived information. The review of sonification of spatial data and interaction strategies with such interfaces is presented in Nasir and Roberts (2007).

The following table lists the most important spatial auditory interfaces described in this subchapter by their project names. It describes also the corresponding interface metaphors and interaction techniques.

Project name and application	Spatial sound configuration and interface metaphor	Interaction technique
AudioStreamer: music browser for personal computer	Three sources at different locations, cocktail party effect	Head movements
Dynamic Soundscape: browsing environment for audio recordings	Sound source orbiting around user's head, audio cursor	Freehand gestures
Audio Hallway: interface for browsing collections of sounds	User travels up and down the audio hallway and enters various rooms	Head movements
Touch Player: interface for controlling a portable music player	Sound sources on horizontal line, stereo panning	Touchscreen gestures
EarPod: portable audio device	Multiple rings with sound sources, stereo panning	Touchpad gestures
Nomadic Radio: wearable messaging application	Spatial sounds with ring metaphor	Speech commands and tactile input
Diary in the Sky: auditory calendar on PDA	Spatial sounds on a clock-like layout surrounding user's head	Touchscreen
AudioGPS: non-visual interaction with GPS	Spatial sound indicating direction of motion and distance to destination	
Audio progress bar	Fixed reference sound source and moving sound source indicating progress	

4.2 Teleconferencing

The increased bandwidth of cellular networks and a variety of available wireless technologies enables mobile device to also engage in teleconferencing and multi-party conversations. A teleconference (Teleconference 2014) is "the live exchange and mass articulation of information among several persons and machines remote from one another but linked by a telecommunications system." Although videoconferencing is becoming more and more popular, audio conferencing still remains an important mechanism for remote collaboration. The audio-related hardware of mobile devices limits participants to mono audio, where multiple sound streams from various people are multiplexed into a single audio output. Most of the work reviewed above demonstrated the higher efficiency and improved perception of multiple sound sources if they originate from different spatial positions. There have also been many attempts to benefit from this artifact in teleconferencing systems. A common feature of the proposed systems is some sort of a virtual 3D acoustic environment where individual participants of a joined telephone call are each placed at a unique spatial location within the virtual world.

One of the earliest attempts at sophisticated teleconferencing system is based on a wearable communication space (Billinghurst et al. 1998). It consists of body-stabilized display that gives the user the impression of being surrounded by a virtual cylinder of visual and auditory information. The main component is a head-mounted display and a head tracker to enable the user to look around the entire space. This system allows remote collaborators to appear as virtual avatars distributed in front of or around the user. Each avatar represents a live video and audio stream, with the audio being spatialized in the correct spatial location. Users have the freedom to turn to the face of a selected collaborator and engage in face-to-face communication while being aware of other conversation in the space. By using a see-through head-mounted display, a user can see also the real world and engage in real-world tasks.

The Vocal Village (Kilgore et al. 2003) system is a similar communication tool which also allows audio conferencing with simple real-time audio spatialization. It uses binaural audio signals that represent individual conference participants at different spatial locations. The evaluation experiments with the system demonstrated a significant improvement of the ability to remember authors of different speech content (e.g., who said what) when participants were able to manipulate individual voices' positions and to move them to arbitrary positions. Spatialization also reduced the perceived difficulty of identifying an individual speaker during the conversation. The main disadvantage of the system reported by the authors was the lack of externalization cues resulting in "within the head" spatialization.

It has been shown that a non-visual virtual communication space for mobile users can also be generated by a simple mobile phone and an inertial tracker (Billinghurst et al. 2007; Deo et al. 2007). The proposed system consists of a mobile phone mockup with a small LCD screen, keypad, stereo headphones, and two tri-axis digital gyroscopes. The mockup device is connected to a personal computer which takes care of the audio and video processing. The gyroscopes enable head and device tracking in order to provide several different audio conditions: mono, fixed spatial and head-tracked spatial binaural, etc. For all spatial audio conditions, the individual sound sources represented remote participants of a teleconference call and were therefore placed in a horizontal circular plane surrounding the user. The head-mounted inertial tracker enabled adaptation of the user's head orientation in the virtual sound space.

Internet protocol (IP)-based networks have also been extensively used to conduct free or very low-cost telephone-like voice conversations. Numerous Voice-over-IP (VoIP 2014) software applications can be found on the market, many being available to users as open source packages. The majority of them support video communication and multiparty conferences and collaborations. Several researchers have therefore proposed extensions of such VoIP systems with 3D sound functionalities. The 3D Telephony System (Hyder et al. 2009, 2010) generates a virtual 3D acoustic environment where individual participants of a joined telephone call are each placed at a unique spatial location within the virtual world. The proposed setup extends the open source internet phone platform called Ekiga (2014) by adding the 3D sound processing software Uni-Verse (Kajastila et al. 2007). The system stores 3D geometric data of a virtual environment, which can be accessed and modified via the UDP-based protocol. It also provides a tool for uploading VML files describing the geometry of virtual rooms. A different virtual room can be used for each call with a different seating plan, a different maximum number of participants and also a different virtual conference table. A user with headphones can identify individual participants by locating them and listening to just one of them, despite multiple talkers speaking simultaneously. The proposed system can also be used in other virtual environments such as gaming and VR simulations.

Another example of the integration of a spatial sound convolution engine into an open source VoIP system, called Mumble (2014), is described in Rothbucher et al. (2010). The Mumble platform was originally developed to support communication without time lag between people playing online games. It has its own positioning module which virtually places users in their avatar positions by simply multiplying the audio signal with an attenuation coefficient. This module has been extended with a 3D sound convolution engine, which uses 3D sound rendering with HRTFs. The incoming audio streams of active conference participants are now filtered with a pair of HRTF filters according to their spatial position. A HRTF dataset can be easily exchanged, allowing users to apply their own individualized HRTFs. The proposed solution was later generalized by extending telephone conference server software with a similar functionality (Rothbucher et al. 2011). This project focuses on compatibility with existing VoIP infrastructures and Session Initiation Protocols (SIP) to support existing telephone systems.

cAR/PE (Regenbrecht et al. 2004) is a more futuristic teleconferencing system which allows three participants at different locations to communicate over a network in a simulated face-to-face environment. The system comprises live video streams of the participants arranged around a virtual table. The meeting takes place in a virtual room which also includes a large 2D virtual presentation screen. Spatial sound is used to indicate different user positions and locations within the room. There are no special hardware requirements for the system. The user sits in front of a standard monitor equipped with one or more cameras, loudspeakers (2.0 or 7.1 configuration) or headphones, and a microphone. It also enables the exchange of electronic documents (e.g., text documents, presentations slides, spreadsheet documents, pictures, etc.).

Different comparisons have been made between centralized (server-based) and client-based filtering of spatialized audio for teleconference systems (Reynolds et al. 2008). When all processing is fully centralized and performed by the server, such a system benefits from larger and more distributed groups of users, but mostly lacks manipulation options. Spatial manipulation is possible only if customized filtering and mixing processes are performed by the server for each user separately. On the other hand, if sound streams are filtered locally on client devices, they can be manipulated by the user in real time, although the processing power is limited due to resource constraints. The authors propose a solution of a distributed processing model, where all available streams can be filtered with a large order of non-individualized HRTFs at the server or filtered locally on each client, if possible. A general model which also aims to support low power terminals is proposed in Reynolds et al. (2009). It does not require a centralized spatial processing functionality and can easily be implemented within IP multimedia subsystems (IMS).

4.3 Virtual Reality (VR) and Augmented Reality (AR)

Virtual Reality (VR) is a computer-generated synthetic environment which can simulate a person's physical presence in different places in the real or an imagined world (Virtual Reality 2014). It can recreate various sensory experiences, including seeing, hearing, feeling or even smelling and tasting. One of the most important features of VR environments is 3D perception, which includes video and sound. Spatial auditory interfaces in VR and mechanisms for the production of realistic acoustic perception are a big challenge and the main point of interest for many researchers. An accurate spatial reproduction of sound can significantly enhance 3D perception, which is important for achieving the sound localization of visual objects. Different issues related to immersive audio systems and their requirements are discussed in Kyriakakis (1998), from fundamental physical limitations to technological drawbacks. The discussion includes identification of fundamental physical limitations important for the performance of immersive audio systems, evaluation of their development, and technological considerations that affect these systems in present and future designs. Similar issues related to auralization, including history, trends, problems and possibilities, are also discussed by Kleiner et al. (1993).

Lentz et al. (2007) describe a system for free sound field modeling and its reproduction in VR systems called VirKopf. It is a software-based solution which runs on a set of dedicated but off-the-shelf machines interconnected by a network. It offers a binaural sound experience without the need to wear headphones. Only four speakers are enough to create a surrounding acoustic virtual environment for a single user. The two most important parts of the system are a geometrical acoustics model and the corresponding algorithms for simulating room impulse responses in real-time applications. The final goal is to simulate a virtual headphone—reproduction of correct binaural signals at the ears—by using crosstalk cancellation mechanisms and without limiting the user to only one sweet spot. The head tracker, based on an optical tracking system, is used for acquiring the current position and orientation of the user. Additionally, a large database of all possible HRTFs is needed. The authors report good overall performance of the system, but also suggest several improvements to the existing algorithms (e.g., consideration of sound insulation and diffraction, optimization of algorithms to increase performance, etc.).

Augmented Reality (AR), on the other hand, combines real-world environments with computer-generated sensory input, including video, sound or GPS data (Augmented Reality 2014). While VR completely immerses a user inside a synthetic environment and prevents him or her from seeing the real world around them, AR allows the user to see the real world and superimposes various virtual objects upon it (Azuma 1997). AR supplements reality instead of replacing it and enables simultaneous interaction with real and virtual objects in the scene. One of the basic, and also most important, problems of AR is registration. Registration means the proper alignment of the real and virtual world to achieve the illusion that they coexist. The registration quality depends on tracking accuracy, which includes technologies such as digital cameras, optical sensors, accelerometers, gyroscopes, GPS units, etc.

For the visual AR it is important to select and use an appropriate display technique. The most common type of display is a Head Mounted Display (HMD) (HMD 2014) which can be a see-through HMD or a closed-view HMD. The first type of display lets the user see the real world directly while the second type requires a camera to record the real world and display it on a screen. The screen is usually a combination of two small separate screens that display different content for each eye. Therefore, it is possible to project 3D visual content.

Audio AR (AAR), on the other hand, superimposes virtual sounds on the natural sounds of the environment (Härmä et al. 2004). The concept was originally proposed by Krueger (1991). One of the earliest designs of AAR is an in-building mobile auditory interface called Audio Aura (Mynatt et al. 1997, 1998). Users wear wireless headphones and small electronic badges which can be tracked and identified by a network of infrared sensors in the building. When a user enters a room or certain area or approaches various physical objects in the room, he or she can hear individualized sound cues such as a summary of the newly-arrived emails, meeting reminders, recent acquisitions, etc. Different environment sounds and metaphors are created in which various sets of functionalities are assigned to various sounds. One such example is the sound of a beach where crying seagulls indicate new emails and the intensity of the sound of the waves indicates the intensity of group activity. The goal of this sound design is to present auditory cues in the background or auditory periphery in order not to draw too much of user's attention. In this case, the auditory cues are not spatialized and the system only supports indirect physical interaction, e.g., a user has to enter a room to activate an audio cue.

The use of spatial audio in AAR was first reported by Cohen et al. (1993). Multidimensional audio windows (MAW) are used to explore the use of binaural signals with virtual spatial sound sources. Spatialization is made by convolving monoaural signals with corresponding left and right HRTF filters. The system is operated through a GUI which enables the user to dynamically select the desired virtual location. As an example of such a location, the authors proposed a place with a virtual phone in an office with the ringing sound coming from the corresponding spatial location. With such system it is possible to provide a telepresence experience through a robot slave operated by a human operator. The operator has a binaural sensation of real and virtual sound sources.

The main idea of the AAR is to correctly superimpose natural (e.g., environment) and virtual (e.g., artificially synthesized) sounds. In a perfect AAR system, a listener should not be able to differentiate between natural and virtual sounds, or there should be a clear distinction between the two—depending on the concrete setup requirements and goals. The problem of the correct registration and alignment of these two types of sound sources has been addressed by Härmä et al. (2004) and Tikander et al. (2003). The Mobile Augmented Reality Audio (MARA) or Wearable Augmented Reality Audio (WARA) (Härmä et al. 2003) is a prototype system of transducer configuration consisting of earplug-type headphones (earphones) with integrated miniature microphones. Microphones are used to record environment sounds and send them directly to the earphones, enabling an authentic perception of the real acoustic environment. The perception of the real world through microphones and earphones differs from real open-air perception due to the frequency characteristics of audio equipment and also the obstruction of the human ear canal. The difference can be minimized by various equalization methods and signal processing techniques. However, if virtual synthetic sounds are superimposed onto the acoustic environment, an augmented audio environment is created. The mixture of both environments should be carefully designed in order to produce a coherent perception for the user and to create a meaningful fusion of the virtual and real world. Virtual audio sources are added to the earphone signals using HRTF directional filters to simulate responses from the spatial location of sources to microphone locations in the ear canals. The location of the user's head represents the origin of a coordinate system of the virtual auditory space. The effectiveness of the MARA system was proved with a listening test which compared virtual and real sound sources, where test subjects were unable to distinguish between the two.

Spatialized AR Audio (SARA) (Martin et al. 2009) is also a system for superposition of virtual sound sources on the real acoustic environment, where virtual sound sources are perceived as coming from different spatial locations. It uses special open and acoustically-transparent earpieces in order to avoid the occlusion effect reported by Härmä et al. (2003) and the use of microphones. Additionally, it includes a set of compensation filters for improving localization in virtual space. The filters are computed for each earphone by taking into account their impulse responses. The proposed system was evaluated in two localization experiments, the first testing the ability of the test subjects to localize real sound sources while using SARA and the second evaluating their ability to localize virtual sound sources created by the SARA system. The results proved the high accuracy and usefulness of the proposed system, although the reported disadvantage of these earphones was their poor low-frequency response.

AAR systems offer a perfect platform for various navigation and orientation services and applications for the selected audio content should be played at the exact location and at the exact moment in time. The LISTEN project (Gerhard 2001) is an information system for intuitive navigation of visually-dominated spaces. It creates a dynamic virtual auditory scene which augments real environment sounds. The main component is a motion-tracked wireless headphone for real-time reproduction of spatial sounds (e.g., speech, music, and sound effects) which is related to the user's current position and orientation. The composed sound stream representing the final audio output of the system depends on the user's spatial behavior, visit history, interests and preferences, etc. The idea is to investigate relationships between visual, auditory, and imaginary spaces. When the user moves through a real space, he or she also automatically navigates the attached complementary acoustic space. Virtual acoustic labels can be attached to any physical object in the real world.

An example of an indoor mobile AAR system is presented by Vazquez-Alvarez et al. (2014). Its goal is to provide eyes-free mobile interaction with location-based information in an indoor art exhibition space. It is based on a complex multilevel spatial auditory display with the simultaneous and sequential presentation of sound elements. The system runs on a mobile phone with an HRTF library, a wireless internet connection, and a pair of headphones. The infrared camera in combination with IR tags and accelerator-based sensors is used to track the user's position and orientation in order to activate the audio interface in the proximity of the selected art pieces. The auditory interface consists of two layers: a top-level sonification layer and a lower level interactive layer. In the sonification layer, a set of chaptering voices are played within the proximity zone, advertising the content of individual art pieces. This layer is played automatically, while the second interactive layer has to be activated by a button press. In the spatial audio version of the interactive layer, a set of three earcons representing different menu items are played in a radial menu around the user's head—to the left, to the right, and in front of the nose. They are played sequentially or simultaneously (depending on the experiment condition) and selected by pressing one of the three buttons on a remote interaction device.

Ec(h)o (Wakkary and Hatala 2007) is a similar and even more sophisticated approach which comprises an audio museum guide and a situated game within the museum. The museum topics are described by audio objects based on the user's level of interest, location, and interaction history. The user's movement within the exhibition space represents the main navigation cue by creating an ambient soundscape related to the artifacts in their proximity. The entire space is divided into several different interactive zones. Interaction with the selected artifact is based on a spatial audio display operated through a tangible interface in the shape of a wooden cube. Different cube faces represent different interface options. The cube rests in the user's hand and can be rotated in any direction. The side of the cube facing upward presents the selected option. Three cube faces are presented with three auditory channels: left, right, and center. The sounds used in the interface are various informal speech recordings. The global aim of the proposed game guide is to improve visitor engagement by including playfulness as an important addition to the functionality of the system.

Several researchers have also proposed very interesting outdoor navigational applications, with the common goal of augmenting the real world with descriptive and informative audio cues. 3DAAR (Sundareswaran et al. 2003) is an example of a wearable AR system intended for cuing a mobile user about objects, events, and potential threats that are outside of his or her field of view. The system is mounted on a vest and customized helmet and consists of a wearable computer, software-based 3D sound library, wireless communication, speech recognition system, GPS location module, VoIP module, magnetometer-based head tracker, and several buttons. A GPS module and a head tracker provide the system with the user's position and orientation, which are then used to present sound cues as a set of 3D audio icons. The icons are played sequentially by azimuth angle. The wireless link is used to broadcast the position of the user to all mobile users. The 3DAAR system was used for various experiments with 3D sound localization and also for determining the role of perceptual feedback training in localization accuracy.

The Sound Garden (Vazquez-Alvarez et al. 2012) project is named after a landscape in the Municipal Gardens in Madeira. It is a mobile AR setup consisting of a mobile phone with 3D positioning library, external GPS receiver, acceleration sensor, and a pair of headphones. It serves as an interactive guide for finding different physical landmarks within the garden. The idea is to use movement sensor data and user feedback to describe the user experience in an exploratory mobile AR environment and to find the most appropriate auditory display configuration. The basic audio content consists of earcons—recordings of animal sounds, such as an owl, goose, cricket, nightingale, or frog—which indicate different regions of the park. The mapping between sounds and landmarks is abstract and the relationship has to be learned. Short synthesized speech clips are used to provide basic factual information about the landmarked sites. When the user enters the proximity zone (a radius of 25 m from the target) the animal sound starts playing from the corresponding direction of the target and with the corresponding loudness. The loudness increases as the user walks closer to the target. When he or she locates the final target (within a radius of 10 m from the target), the speech audio clip provides descriptive information about the target.

Audio Stickies (Langlotz et al. 2013) is another outdoor AR platform that provides spatial audio augmentation on simple mobile devices. The "stickies" are spatial audio annotations generated by the users and linked to physical objects in the environment. They try to simulate well-known and widely-used written sticky notes. Their position in the environment is indicated by visual hints on a real-time video recorded by a mobile phone. Users browse the environment through a display and perceive an AR view of the surroundings. Only the stickies in the center of the image are played and the selection is made by moving the phone and changing the displayed image. Several stickies can be played simultaneously if they are all placed in near proximity. In order to attach the audio annotation to the real object a user has to select the object by touching it on the display. The spatialization of the sound sources is based mainly on stereo panning techniques due to limited capabilities of smart phones.

Several authors have proposed the use of VR and AR technology as a platform for the reproduction and manipulation of music in real and virtual environments. Haller et al. (2002) and Dobler et al. (2002) reported an interesting AR interface for real-time positioning of virtual musical instruments in real space. Users can see, listen to, and also feel various musical sound sources in space and move them around with the aid of a special pen. Sound sources are visualized as 3D models of musical instruments and their sound is localized to originate from the corresponding spatial position. Two examples of similar AR interfaces for manipulating spatial sound are AR/DJ (Stampfl 2003a) and 3deSoundBox (Stampfl 2003b). AR/DJ is a simple application which allows two music DJs in a club to play manipulate live sound sources through an AR interface and place them anywhere in 3D space in the club. Sound sources are visualized in a 3D model of the real space and can be moved using a special pen with visual tracking markers. 3deSoundBox, on the other hand, is an acoustic platform-independent component for using and exploring 3D sound in AR applications. It can work with any number of speakers and has a very scalable architecture.

Digital Interactive Virtual Acoustics (DIVA) (Savioja et al. 1999) is another real-time environment for a full audiovisual experience, such as a virtual music concert. It is based on signal processing techniques for spatialized reproduction of sound sources. There is support for two simultaneous users, one representing a conductor and the other representing the listener. The conductor is tracked in the space and controls the orchestra (e.g., tempo and loudness) with different movements. The tracking is based on a coat with magnetic sensors. The orchestra can consist of real and virtual musicians. Virtual musicians are presented with animated human models and their positions are manipulated through a GUI. The listener can also move around the virtual concert hall, if required. The final sound output is reproduced through headphones or loudspeakers. The room acoustics are simulated using a time-domain hybrid model that takes into account the direct sound and early reflections. These are spatialized using HRTFs or by simple approximation based on adjusting only ITD and ILD. The technology used in the DIVA system can also be used in other applications (e.g., games, multimedia, virtual reality, etc.).

The AR environment and simultaneous presentation of visual and audio cues also offers an ideal tool for studying the human ability to localize real and virtual sound sources. Zahorik et al. (2001) reported on localization experiments with non-individualized HRTFs and low-cost 3D audio equipment in which he wanted to find out if perceptual training can reduce localization errors. They used a head-mounted display for the simultaneous presentation of visual and auditory cues in an AR environment. Another study on the impact of 3D sound on perception in an AR environment was reported by Zhou et al. (2004). The effects were studied in two different scenarios, combining visual and auditory cues as well as a complex game-based environment. The results of the study demonstrated the large importance and impact of 3D sound on significant improvements in depth perception, task performance, and human presence and collaboration. Mariette (2006) proposed a methodology for studying sound localization outdoors when listeners are mobile and virtual sound sources can correspond to real visual objects. The selection of the target was performed by physically walking to the destination, addressing the problem of multimodal perception and subject interaction via self-motion. The goal was to study the effect of HRIR filter length and the final quality of binaural rendering on localization errors in such multimodal perception. The wearable AR systems 3DAAR (Sundareswaran et al. 2003) and WARA (Härmä et al. 2003) described above were also used to observe the role of feedback training on sound localization accuracy. Sodnik et al. (2006) reported on a sound localization experiment in a tabletop AR environment. The AR scene consisted of a large number of identical visual objects distributed in the virtual space. Any object in the scene could be sonified with a spatial sound source, but only one at the time. Users were asked to localize the "noisy" model by moving through a virtual AR scene wearing a head-mounted display and stereo headphones, as shown in Fig. 4.2. The results demonstrated the human ability to localize the azimuth of sound sources with a higher accuracy than elevation or distance.

4.4 Aircraft

One of the crucial components in the cockpit of a modern aircraft is the Traffic Alert and Collision Avoidance System (TCAS), which is used to alert the flight crew about other aircraft in close proximity (TCAS 2014). The TCAS system works independently of air traffic control and monitors the airspace around an aircraft for other potential aircraft which are equipped with an active transponder. It is usually integrated into the navigation display or electronic horizontal situation indicator and

Fig. 4.2   User performing localization test in an AR environment (Sodnik et al. 2006)

it uses auditory and visual displays for interaction with the pilot (Begault 1993). In most situations the visual display provides crucial information about the surrounding aircraft, while the auditory display is used only as a redundant warning system. There are three types of alerts: traffic advisory, resolution of advisory, and clear of conflict (TCAS 2014). The first alert is an informational message only displayed on the visual display that serves situational awareness. The second alert also includes a spoken auditory message and arises when there is a potential collision within the next 40 s. The third alert requires the immediate reaction of the pilot and consists of a visual and spoken auditory message with instructions on an evasive maneuver. It occurs approx. 25 s before a potential collision.

Begault (1993) proposed the use of a spatial sound in a head-up auditory display to provide the spatial information that is usually shown on the head-down visual display. The spatial sound indicated an out-the-window target direction relative to a forward head position. The sound sources were virtualized in the azimuth plane using HRTF filters and played through headphones. The potential head movements were not correlated with spatial auditory output. The proposed system was evaluated in an experiment with trained crew members in a flight simulator who were asked to start searching for the targeted aircraft based on auditory messages and to locate it visually out of the window. The target acquisition time was measured in spatial and non-spatial (using only one earpiece) experiment conditions. The search procedure was considered successful if the target was acquired within 15 s. The presence of spatial cues in auditory warnings significantly reduced the visual search time (by approx. 2 s compared with non-spatial cues).

The use of spatial auditory warnings within TCAS also represents a great advantage over standard visual-audio TCAS (visual information + mono audio) warnings

(Begault and Pittman 1996). The target acquisition time in this case can also be reduced significantly. The use of spatial audio warnings with visual TCAS results in approx. 12 % faster response time compared to just mono sound (Oving et al. 2004). Even better performance can be achieved with a combination of spatial sound and verbal directional information on required safety actions.

A similar experiment in the military domain is described by Bronkhorst et al. (1996). In this case, spatial audio is used to report the location of a target in a jet fighter. A group of pilots in a simulator was asked to locate and intercept an enemy jet based on visual 2D radar displayed on a head-down display in front of the pilot. In the second condition a 3D auditory display was used for the same task. The combination of both solutions seems to be the fastest interface for the selected task, while there is no significant difference when using just one of them independently.

In follow-up studies, the effect of two simultaneous auditory warnings related to two different tasks has also been studied (Veltman et al. 2004). The experiment consisted of groups of primary and secondary tasks performed with a visual Head-Up Display (HUD) or Head-Down Display (HDD). Primary tasks were always performed one at the time, while secondary tasks could also be combined and performed in parallel. In the primary tasks the pilots had to locate and pursue a target that appeared somewhere around the jet and was signaled on the HUD. Successful capture was indicated by a color signal on the HUD and non-localized auditory signal. In the secondary tasks the pilots were asked to pay attention to some visual information on the HUD or HDD and react in certain situations. In some experiment conditions, 3D audio was added to the HUD information in the primary task and sometimes also to the HDD information in the secondary tasks (in these conditions two sound sources were presented simultaneously). The greatest benefit of 3D audio was shown when it was combined with the HDD. In these cases, the frequency of vertical eye movements and the perceived cognitive workload were reduced significantly. The superposition of 3D audio on the HUD did not improve performance considerably. Another attempt at supplementing the HDD with high-fidelity 3D audio was also reported by Parker et al. (2004). In their experiment, the participants were asked to acquire the image of a target aircraft on an HDD. When combined with 3D audio cues, the acquisition time was faster, the perceived workload was lower, and situational awareness was improved.

Nowadays, Unmanned Aerial Vehicles (UAVs) also play an important role in various military operations (Air 2004). They are particularly suited to complex and dynamic operations where the utilization of traditional manned aircrafts would be very dangerous or even impossible. Tele-operation of such UAVs is very different from piloting an actual aircraft and incorporates manual and automatic controls. The operator interfaces emphasize the presentation of information through a visual display, often risking the overload of the visual information processing capacity of the operator. Therefore spatial auditory display technologies represent great potential as isolated communication channels or as components of a multimodal display system. They should be spatially, spectrally, and temporally manipulated, and thus lead to a strong sense of presence. Information about the current and future states of a complex and highly-dimensional environment could be presented in a very intuitive way. Auditory motion perception can, for example, be used to report on the trajectories of various elements in the environment. Spatial audio displays can lead to a high level of realism and task engagement, improving overall operator performance.

Helicopters, on the other hand, are significantly different types of machines (Houtsma 2003). The SPL levels inside the cockpit can often be over 100 dB, requiring the pilots to wear some sort of ear protection. The most common solution is a pair of helmet-mounted earplugs which offer protection against noise and enable clear electronic communication. However, the ear protection attenuates all types of sounds, including potential auditory warning signals produced by the machine. Each type of helicopter has a unique warning signal system comprising of speech, non-speech, and mixed messages. Spatial sound has also been used in a warning display for system malfunctions in helicopters. Haas (1998) used the location of virtual spatial sounds to report the actual location of a system malfunction in the helicopter or to report the location of a visual warning indicator within the cockpit. Spatial sound cues resulted in approx. 28 % faster response times compared with solely visual warning systems.

4.5 Vehicles

Similar to aircraft cockpits, different Collision Avoidance Systems (CASs) are also used in vehicles. They are used to detect other vehicles and obstacles on the road and to warn the driver if the distance or speed of approach exceeds a certain threshold. Graham (1999) proposed and tested the use of auditory icons for these warnings, since the visual sense is highly occupied during driving tasks. The auditory messages should not be confusable with environmental sounds, such as engine noise, indicator signals, seatbelt warnings, etc. They should also be very intuitive and correctly interpreted without prior instructions and learning. They should provoke fast and accurate response from the driver (e.g., emergency braking, steering maneuvers, etc.). In his experiment, Graham compared the proposed auditory icons to conventional speech warnings and to abstract non-speech sounds in terms of reaction time, number of inappropriate responses, and subjective ratings. The auditory icons actually proved to provoke significantly faster response times, although they also provoked a larger number of wrong responses. Graham suggested the auditory icons have many advantages over conventional sounds in terms of response time and subjective ratings, but they could be further improved by the use of spatial sounds (e.g., direction of sound could indicate the direction of danger) and other manipulations of sound signals.

The Advanced Driver Assistance System (ASAD) in vehicles provides automatic lateral and longitudinal control, reversing and parking aid, etc. These systems also require an effective and intuitive user interface in order to inform the driver about current activities and to provoke immediate reaction when required. Bellotti et al. (2002) explored the use of 3D auditory interfaces for providing drivers with information from the ADAS system. According to their hypothesis the use of 3D sound should reduce eyes-off-the-road time and reduce the number of false alarms. They proposed assistive systems for providing lateral control, longitudinal control, and support for parking. 3D sound techniques are used to generate sounds that correspond to objects detected as potential hazards and obstacles (e.g., lateral lanes, guard-rails, other vehicles in front or behind, cars in the blind spots of the lateral mirrors, etc.). The 3D sound scene was created by four speakers: two in the front of the car and two in the rear. Tones masked with a square wave were used as alarm signals to enable fast and accurate localization. The system proved to be effective at detecting and localizing potential collision situations around the car, although some test subjects found the system annoying. The latter could be improved by designing the sound messages better. The 3D sound messages could also be integrated with other interface modalities (e.g., visual alarms).

Suzuki and Jansson (2003) also studied the use of spatial sound cues to warn the driver about line deviations. They compared four types of warning cues: monoaural beeps, stereo beeps, steering vibration, and pulse-like steering torque. The stereo beeps originated from the direction of the potential line crossing and were intended to warn the driver to move back in the opposite direction. In the so-called predicted scenario, the users were informed about the line deviation warning system and its functionality, while in the unpredicted scenario they were exposed to the signals unexpectedly. The steering vibration signals were reported to be the most efficient interface for the unpredicted scenarios, while the stereo beeps proved to be better when test subjects were aware of its meaning and functionality. The authors also reported several cases of misinterpretations of the warning signals where users turned the steering wheel in the wrong direction. However, none of these cases happened in the auditory conditions.

Ho and Spence (2005) conducted a series of experiments in which they observed the efficiency of localized auditory cues for warning the driver about approaching vehicles and other critical driving events. In half of the cases, the sound warnings were localized to indicate the direction of the danger, while in the other half the sounds were not localized. In the third part of the experiment, the warning sound cues were wrongly localized and therefore indicated the wrong direction of the danger. Drivers reacted more rapidly to dangerous events seen in the rearview mirror when warning sound cues were presented from the same direction. On the other hand, the authors reported only marginally faster responses for dangerous events seen in front through the windshield. When comparing non-localized or wrongly-localized sound to correctly-localized cues, non-localized or wrongly-localized sound was shown to decrease performance, possibly due to extra processing time. The authors concluded that invalid spatial warning cues can impair performance, while correct warning cues can facilitate it significantly.

In another experiment, the same authors (Ho and Spence 2009) also assessed the relative speed with which people reacted to auditory warnings (white noise bursts) originating from the back of the head (at a distance of 40 cm) or from far in front of the head (at a distance of 80 cm). Both types of auditory cues were also compared to vibrotactile warnings presented to their waist or wrist. The auditory signals presented close to the user's head (within the so called "peripersonal space") proved to be the most efficient warning cues and facilitated the fastest head-turning response. They also proved to offer an effective means of alerting a driver to orient his or her eye gaze toward locations of interest and danger. The auditory warnings presented in close proximity to the driver's head significantly outperformed the other two groups of warning cues in terms of latency of response and at the level of the participants' visual discrimination performance.

Current vehicular research focuses also on the development of autonomous and semi-autonomous vehicles. Automation of the primary driving task can significantly reduce the mental workload of drivers but can also diminish their sense of control and the feedback about the state of the vehicle and its basic driving properties: steering, gear changing, acceleration, braking, etc. Beattie et al. (2014) report on a driving simulator study in which they attempted to determine the suitability and effectiveness of a spatial auditory display for re-establishing the afore-mentioned sense of control to drivers. They also observed whether spatially positioned sound sources could enhance driver awareness of the intended actions of autonomous vehicles. They recorded realistic sounds of a vehicle (e.g. acceleration, braking, gear changing, etc.) and played them as spatial and non-spatial sounds in a driving simulator. The spatialised auditory presentations of these sounds were reported to be very effective at alerting drivers about the intended actions of the autonomous vehicles and also for re-establishing the sense of control to drivers.

On-board entertainment displays or In-Vehicle Information Systems (IVIS) also require user interaction and advanced user interfaces. Sodnik et al. (2008) describe a 3D auditory interface based on a ring metaphor which represented a hierarchical menu structure. Individual menu items and commands are presented as virtual sound sources placed on a circle surrounding the user's head (Fig. 4.3). The sound sources are spoken words—readings of the menu. There are between one and six sources (e.g., depending on the experiment condition) played simultaneously that represent currently available menu items and that enable the user to select any of them. When a user selects one of the items, a new set of items, which represent the corresponding submenu, is loaded. In addition, non-localized background music is used to indicate an individual branch of the hierarchical menu. Its central pitch changes according to the current depth of the user in the submenu (the pitch lowers when the user moves deeper in the menu and vice versa when he or she moves back up in the hierarchy).

The interaction with the interface is conducted through a custom-made device that consists of a scrolling wheel and two buttons. It is attached to the steering wheel so that it may also be easily utilized while driving. The proposed auditory interface was evaluated in a user study where it was compared to typical HDD that displayed the same menu structure. It proved to be equally fast as a visual display, while also being much safer and less cognitively demanding. The latter was reflected in the lower number of errors made when driving and performing secondary tasks with both displays. On the other hand, the use of simultaneous spatial sound did not prove to have any significant advantage over a simple mono sound representing one selected source (e.g., menu item) at a time. The majority of test subjects reported the multiple simultaneous sounds to be hard to perceive and to understand while driving.

Fig. 4.3 The virtual ring
of sound sources
distributed equally around
the user's head (Sodnik
et al. 2008)

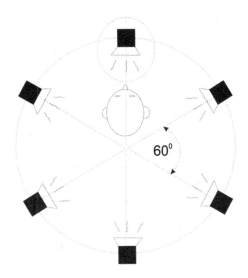

The use of speech-based auditory menus for IVIS has also been studied by Jeon et al. (2009). They investigated the advantages and disadvantages of enhanced auditory cues (e.g., spearcons and spindex) in typical in-vehicle scenarios compared with normal text-to-speech cues. The experiment also showed that an enhanced auditory interface improved the user's performance and safety when utilizing the IVIS in the presence of a visually-demanding primary task.

In the field of visual representation of information in vehicles, the HUD is the current state-of-the-art solution for reducing driver errors that originate from distracting interfaces. The major advantage of HUDs compared with HDDs is supposed to be the response time to unanticipated road events. A typical HUD displays screen-fixed icons indicating current speed and some basic navigational cues. With the aid of eye trackers, these displays can also be used for 3D AR visualization. Tonnis and Klinker (2006) proposed an egocentric AR perspective where special navigational and assistance cues are projected in front of the car in the position of the car's front bumper. In addition, a 3D warning sound is also generated from the direction of the potential dangerous situation. A multimodal extension with 3D-encoded sound supports basic acoustic guidance. In such cases, sounds are always hybrids of location and direction indication. The warning system is composed of various harmonic frequencies supporting better localization and indicating the urgency of various warnings. Multimodal warning presentations that combine sound and visual schemes have proved to be an important source of improvement compared with solely visual representations in 2D or 3D.

Jakus et al. (2015) describes a multimodal IVIS device combining HUD and auditory interfaces. Similar to the IVIS proposed by Sodnik et al. (2008), the structure is presented as a hierarchical menu consisting of up to six menu items at each level. It is also operated through a similar interaction device attached to the steering wheel. In three different experiment conditions, the system was displayed on a

HUD, played through an auditory interface, or presented through a simultaneous combination of both interfaces. The use of the HUD showed a significant increase in safety and decrease in cognitive workload compared to the HDD. The majority of users clearly chose the multimodal display as the preferred technology, as it enabled them to change communication modality at will. They could operate the IVIS through a solely auditory menu when driving in dynamic and visually demanding traffic situations and operating the menu through the HUD while stationary (e.g., waiting at traffic lights) or in low-density traffic. On the other hand, no objectively-measured differences were reported between different interfaces (e.g., task completion times, driving penalty points, cognitive workload, etc.).

The evaluation of novel technologies and such novel HMIs described above is often performed in a driving simulator which provides a safe and controllable environment. The final goal of the simulator is to immerse the driver in a realistic virtual world that imitates real-world scenarios and situations. An important component of modern driving simulators is also 3D sound, which can be played through headphones or external loudspeakers. The sound system of the VIRTTEX driving simulator uses binaural recordings of a real vehicle which were recorded with a binaural head in the driver's seat (Blommer and Greenberg 2003). The playback of these sounds on external speakers requires various equalization techniques, which are achieved solely with conventional "off-the-shelf" audio hardware. In another version of driving simulator, Mourant and Refsland (2003) also modeled sounds of squealing tires, the engine noise of autonomous vehicles, and the siren of a police vehicle.

4.6 Visually-Impaired and Blind Users

Visually-impaired and blind users perceive the surrounding world differently to sighted people. They are unable to use their visual sense for orientation, navigation, or mobility and have to learn to exploit other senses such as feeling and hearing. Additionally, their interaction with information and communication technology has to be adapted to use these alternative senses. Spatial sound offers an excellent upgrade of various auditory interfaces for the visually impaired and blind by describing the locations of real world objects and interface modules with positions of virtual sound sources. In this chapter we summarize several different spatial auditory interfaces for the navigation and orientation of visually-impaired users in space, as well as interfaces for computers and mobile devices.

The vOiCe is an early, although still in use, example of a navigational tool for the visually impaired based on the direct conversion of a video image into a 3D audio signal (Meijer 1992). The video signal is acquired by a video camera and converted into a black-and-white image consisting of 64×64 pixels. The image is then sonified with an acoustics scanner moving through all columns of the image from left to right. The output sound image is spatialized so that the direction of the sonar can be clearly perceived by the listener. Individual columns of the image consisting of 64 pixels are represented as 64 harmonics of different frequencies. The pixels at higher physical positions in the image are coded with high frequencies, while pixels at lower positions are coded with lower frequencies. The color, or the intensity of each pixel (from white to black), is coded as an intensity of sound for that pixel. Such an artificial coding scheme has to be learned by users in order to be used effectively and in the real world. The system runs on various mobile devices and can be used as a supportive navigation tool. The authors reported the high usability of the system and the capability of their users to actually "see" the environment with sound.

The majority of modern HCI is based on GUIs consisting of windows, icons, menus, pointers, etc. These are not directly useful for visually-impaired and blind computer users, as they are usually shown on displays and can only be perceived through the visual channel and visual communication. Visually-impaired and blind computer users have learned to use such interfaces with speech synthesizers and screen-reading software—often referred to as Screen Readers (2014). This software provides verbalization of the graphical interaction and enables the user to review the content of the screen and track changes on the screen through voice messages and descriptions. The output is mainly a text-to-speech auditory signal with varying speech timbre, pitch, speed, etc. Another alternative is a Refreshable Braille Display (2014), which represents a selected portion of the GUI and is read actively with the fingers. Both the screen reader and the Braille displays represent just one selected node of the visual interface. The selection is usually performed with a classical computer keyboard.

Crispien et al. (1994) proposed a spatial auditory representation of the computer screen. Screen objects are described as auditory objects consisting of specific sound representations at different spatial positions. The positions of auditory objects are described as three-component vectors in space. 2D rectangular screen positions are transformed into 3D auditory positions in the frontal hemisphere of the listener. The back hemisphere could also be used for the presentation of extraordinary events, such as warning messages and help information. The synthesized speech signals are therefore processed spatially to describe their original screen positions. This can provide immediate understanding of the text layout within a word processor application.

A similar solution of spatially-positioned synthesized speech has also been proposed by Sodnik et al. (2012). A standard screen reader is upgraded with spatial sound and the capability of manipulating and changing the spatial position of the synthesized speech. The system consists of standardized and generally-available software components and a custom Java application (Sodnik and Tomažič 2011). It is integrated into a word processing application that enables manipulation of the spatial positions of synthesized speech in real time. The speech output can be positioned at different spatial locations by varying the azimuth and elevation from $-20°$ to $20°$, respectively. In the case of tables, the dimensions and locations of individual cells are also described as sources at different positions. Text styles (i.e., italic, bold, or underlined text) are presented with different variations of pitch and rate of speech. The proposed system was evaluated in a user study with visually-impaired and blind users and compared to a conventional screen reader equipped with a Braille display. Users were asked to read and process a text file and collect information about its form and structure. The text file included a varying number of columns, text styles, and tables. The proposed spatial screen reader proved to be as accurate as the tactile interface, but more than 160 % faster. It also proved to be very intuitive and well understood by the majority of users.

Mereu and Kazman (1996) also studied the effects of sound in a 3D computer interface with visually-impaired users by asking them to locate a target at a specific spatial position by moving a mouse cursor to that location (up/down arrow keys were used in addition to the mouse to control movement in the third dimension). Test subjects continually hear the target's location as well as the current location of the mouse cursor. The authors proposed a custom sound coding to describe the spatial locations of various objects. The azimuth is coded as a balance between the left and right channel, the elevation is coded as a corresponding change of pitch, while the distance is coded as a variation of the volume of the sound signal. Two different sound signals are used: a simple tone and a musical piece. In another alternative (i.e., the "musical" condition), different musical instruments are used to indicate different spatial positions. When the mouse cursor is moved to a selected spatial position, the corresponding musical instrument starts playing loudly while other instruments are decreased in volume to be heard only in the background. The use of only simple tones proved to be the preferred method for visually-impaired users.

Various binaural simulation techniques and room acoustics can also be used to generate various general interfaces for visually-impaired people, enabling them to use computers and selected tools and applications. Frauenberger and Noistering (2003) proposed a Virtual Audio Reality (VAR) system based on a surrounding spatial audio interface. The windows, menus, and icons of the interface are presented as earcons located at different locations in a virtual room. Various earcons can also be grouped to parent earcons to provide a smaller number of independent sources. The user acts as a pointer device which can virtually move to a desired earcon and interact with it. The interaction is performed with a joystick that has three axes, enabling transversal movements in the X and Y directions and rotations around the Z axis. A head-tracking device is used in addition to detect head rotations and improve localization. The system runs with the aid of an external DSP controlled by a hosting personal computer. The DSP provides the high computational power required for real-time rendering of spatial audio.

Frauenberger et al. (2004a, b) also proposed a spatial auditory interface for navigation in a structured menu, text input, selection of item in a list, selections, and confirmation of various messages and alerts. It represents an auditory version of a graphical interface consisting of menus, lists, text fields, buttons, and check boxes. The metaphor used is a large virtual room where up to six presented items are arranged in a semicircle in front of the user. The interaction with the items is performed through keyboard and head movements. An individual item is played when faced by the user. The active angle of each item is 10°. Additionally, acoustic zooming is provided by weighting the input gain of selected sources. Individual items are presented with auditory icons playing in a loop. Each source position is superimposed by a continuous musical tone whose pitch increases from left to right and, therefore, improves localization. This system is implemented on a customized personal computer equipped with headphones and a head-tracking device. It was also evaluated in a study with four users whom were blind or visually impaired. They were asked to perform different selecting tasks and tasks for entering data into the system. The results showed that complex applications can be effectively used through such a spatial auditory interface, yet interestingly no significant difference was observed between the performance of visually-impaired and normal-sighted users. The authors established that auditory interfaces should be designed without visual concepts in mind (Frauenberger et al. 2005; Frauenberger and Stockman 2006). They introduced a set of mode-independent interaction patterns along with their transformation in the auditory domain. The auditory interface for MS Explorer was chosen as an example in order to demonstrate some problems with the representation of sound.

Sodnik et al. (2011) presented another auditory presentation of the hierarchical menu structure of a MS Windows-like application for visually-impaired and blind users. Individual menu items are presented with prerecorded spoken commands, which are played with different speeds and sound something like spearcons, but slower and still completely intelligible. On each level of the menu, a group of items is presented horizontally or vertically, enabling the user to select any of them (Fig. 4.4).

Each branch of the menu has corresponding background music whose pitch changes with depth (i.e., similar to the mechanism described in Sodnik et al. (2008)). The changing pitch intends to give additional feedback about the current position in the hierarchical structure. In the horizontal version of the interface, the menu items are positioned on a virtual ring surrounding the user's head. Up to three sources are played semi-simultaneously (e.g., one shortly after the other with 200 ms of silence in between). In the vertical interface, the menu items are aligned vertically one above or below the other. In order to improve the localization accuracy of sources at different elevations, an artificial coding of elevation is performed, as suggested by Susnik et al. (2008). This method suggests that elevation of sound sources can additionally be emphasized by variance in pitch. There is a 15 % difference in pitch

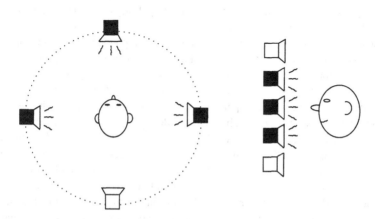

Fig. 4.4 Horizontal and vertical configuration of simultaneous sound sources in the hierarchical menu (Sodnik et al. 2012)

between two neighboring vertical menu items. In both interfaces the interaction is based on a standard keyboard using arrow keys to move in various directions in the menu. The system can run on a normal off-the-shelf computer with no additional specific hardware. It was evaluated in a user study with 12 blind and visually-impaired computer users, who preferred such auditory interfaces with just one sound being played at a given time. The use of multiple simultaneous sounds proved to be too cognitively demanding, resulting in longer task completion times.

Crispien et al. (1996) also investigated auditory interfaces for hierarchical navigation schemes and proposed another version of the "ring" metaphor. The items available at a specific level of the menu form a virtual ring in a horizontal plain surrounding the user. Up to 12 objects are available at each level, with a spacing of 30°. Only three objects in the focus area (e.g., an angle of 90°) are played simultaneously. The focus area changes based on the user's head rotations and rotations of the virtual ring through various input commands. The items which fall out of the focus area fade out and new items fade in. The input commands include 3D pointing, various hand gestures, and speech input. The system requires several PCs as well as DSP units to run in real time.

Jagdish and Gupta (2008) presented a different approach to the auditory presentation of a graphical interface. The system, called SonicGrid, is intended to help visually-impaired and blind users to navigate in GUI-based environments in a non-linear spatial presentation. It is based on an interactive grid layer on the screen that gives non-speech sound feedback. A stereo signal with changing volume between the left and right speaker is used to code the horizontal direction of auditory elements. The mechanism of changing the pitch is used for the vertical dimension. Auditory elements at lower physical positions are presented with low-pitch signals and elements at higher positions with high-pitch signals. The authors evaluated the system and reported it to be effective, but also reported that it required a certain amount of learning time before actual use.

Nowadays, the use of Internet and the World Wide Web services has increased dramatically as more and more services are available solely online, targeting users of all ages and with only basic technical skills. The basic mechanism for defining and describing web pages is the Hyper Text Markup Language (HTML) which uses markup tags to describe their structural semantics. Sighted users access web pages with the aid of visual web browsers which render all graphical and multimedia HTML content. The structure of the document is of key importance to its comprehension. It is even more important when an HTML document is accessed by the screen reading software and audio browsers available to visually-impaired and blind people. Goose and Möller (1999) proposed a spatial audio browser which enables the sequential acoustic projection of a structured hypermedia document to the "Stage-Arc". The latter is represented as a reduced semicircle in front of the listener from which the document content is spoken. Three high volume male voices read the document content out loud (e.g., headers, content, and hypermedia link anchors respectively) from a fixed position at the midpoint of the arc, while a second, low-volume female voice moves along the arc and periodically announces the current position through the document as a percentage. Additionally, when hypermedia links are detected in the content, earcons are played at the corresponding positions of the circle. Various sound effects indicate intra-document links and movements within the document. A 3D "sound survey" is used to indicate the presence of links in the surrounding area of the user's current position within the document. It consists of four distinct earcons reporting on positions and context information in the selected region of the content. The authors report that these features can improve the user's comprehension of HTML document structure and enable effective navigation and orientation within the document.

Another approach of presenting web content aurally has been proposed by Roth et al. (1999, 2000). In the proposed solution, called AB-Web, a macro-analysis provides a spatial representation and exploration of HTML documents, while a micro-analysis can be used to retrieve the content of selected elements. The macro-analysis is performed by moving a finger on a touch-sensitive screen while the system responds with spatial audio feedback about the location and type of content. Earcons are used to represent HTML elements (e.g., a typewriter sound represents text, while a camera shutter sound represents an image) and placed in the 3D sound space to inform the user about the location of these elements. The micro-analysis uses speech synthesis to present selected portions of the document.

Children and adolescents also represent a large part of the blind and visually-impaired population. They need to learn how to use their auditory channel to sense and interpret the environment and their surroundings. Sánchez and Sáenz (2006) focused on children with visual impairment and proposed a 3D virtual environment called AudioChile which uses spatial audio to develop children's problem-solving skills (e.g., searching, mobilization, localization, orientation, etc.). The main components of the virtual environment are models of three major Chilean cities, which are presented through spatial sound and provide a sense of spatiality and immersion in the game. Sound emerges from different obstacles, boundaries of paths and labyrinths, other characters, etc. The authors reported very encouraging and positive feedback from the participants in a user study. In addition, they also proposed AudioMetro (Sánchez and Sáenz 2010), another virtual environment for planning, practicing, and learning about the underground Metro system. It is intended for blind children to help them learn how to navigate and orient themselves in the Santiago Metro Network system and to plan and simulate trips within the city. The idea is that later the users will be able to use this knowledge in a real environment. The sounds used in the virtual environment are representations of the real Metro system, imitating the sequence and duration of the real sounds that correspond to certain situations they will encounter during the real trip. Spatial sounds coming from different directions help users with orientation within the environment. The user study confirmed the positive impact of the software on the development of cognitive, sensory, and psychomotor skills for the orientation and mobility of blind users.

4.7 Brain-Computer Interfaces

Human brains are filled with neurons connected to one another by dendrites and axons (Grabianowski 2007). Every time we think, feel, try to move or remember something, our neurons exchange small electric signals. These signals are generated by the difference in electric potential carried by ions on the membrane of each neuron. The electroencephalographic activity can provide a new non-muscular channel for communication with the external world (Wolpaw et al. 2002). A brain-computer interface (BCI) can therefore be best described as direct communication between the brain and some sort of external device.

BCIs are mostly used for assisting basic human cognitive or sensory motor functions or for augmenting them with additional features and capabilities. They can, for example, use visual stimuli to maintain communication in severely paralyzed patients but only if their vision is intact. Nijboer et al. (2008) proposed the use of sensorimotor rhythm (SMR) as an auditory input signal for BCI. They compared the performance based on auditory and visual stimuli and tried to determine if healthy participants can achieve the same level of performance with both auditory and visual feedbacks. Their tests showed that, at the beginning of training, visual feedback of SMR led to better average BCI performance than auditory feedback; after this initial period, however, both dropped significantly. After several sessions both stimuli performed with comparable accuracy, which indicated that auditory feedback is similarly effective but requires more training.

Various studies of human brain activity concentrate on a PCI300 wave, an event related potential (ERP) evoked in the process of thinking and decision making (Chapman and Bragdon 1964). It can be detected and recorded by an electroencephalogram (EEG) as a positive deflection in voltage with a latency of 200–700 ms after stimulus onset. In the context of BCI, the visual P300 ERP can be used to select letters displayed on a computer monitor in the form of a 6×6 matrix of characters (Farwell and Donchin 1988). The user has to focus on one of the 36 characters in the matrix. The rows and columns of the matrix are then illuminated in random order. When the illuminated row or column contains the chosen letter, a P300 ERP appears. A special classifier is then used to determine the exact character based on the cross-sections of rows and columns. Furdea et al. (2009) proposed the use of ERP elicited by an auditory stimulus as an input signal. They report on a study where they attempted to determine whether BCI users could select characters from an auditorily presented 5×5 matrix of letters. Each character's position in the matrix was determined by two auditorily presented number words indicating the row and the column. The user had to listen to two target sounds representing the coordinates of the selected character. Visual presentation of the matrix was used solely to help the users remember the coordinates of individual letters. ERP speller proved to be quite efficient—in some cases the users were 100 % accurate—but did not achieve the performance comparable to a visual ERP speller in terms of speed and efficiency. Chang et al. (2013) report an extensive study with a spatial auditory speller where visual letters and synthetic vowels were represented as spatial sounds. The users were played visual, auditory or audiovisual stimuli from five different spatial locations. The visual representation was displayed on a large computer display positioned in front of the user, while the auditory stimuli were produced by vector based virtual sound panning technique. The set of spatial positions included the following azimuths: $-45°$, -22.5, $0°$, $22.5°$ and $45°$. The visual and audiovisual spellers outperformed the spatial auditory speller significantly, since the latter resulted in only 52 % mean accuracy of recognition. Nakaizumi et al. (2014) proposed an extended version of the spatial auditory BCI speller by using HRIR for virtual sound image spatialization with headphone-based reproduction. The monaural sounds of vowels (a, i, u, e, o) were spatialized with filters from the CIPIC HRTF library and positioned at azimuth locations of $-80°$, $-40°$, $0°$, $40°$ and $80°$. The EEG results confirmed the P300 responses of users. Also in this case the mean accuracy was not very good but showed the tendency towards improvement with more training.

The majority of auditory BCI systems results in a binary output where ERPs are triggered by two sequences of target and non-target tones. In the experiment reported by Hill et al. (2004) both ears received the same sequence simultaneously but with a different inter-stimulus interval. The users were asked to focus on one of the streams by counting the number of target tones in that stream. Due to the difference in inter-stimulus interval, the ERP response to the left channel differed from the response to the right channel. Similarly Kanoh et al. (2008) explored the human capability of separating audio streams to output a binary BCI command. Two different streams were played to the user and if the inter-stimulus interval of the streams was short, the user could naturally segregate them into two independent streams that could further be identified as two different ERPs. Schreuder et al. (2010) suggested the use of spatial sound and spatial cues as stimuli in BCI, since this could help the users to discriminate more than just two different classes. They presented the users with five speakers positioned at different locations in front of them. The speakers were at ear height and at a distance of approx. 1 m. All speakers were used to play the same stimulus—a 40 ms sequence of white noise with a tone overlay. The discriminating cues in the stimuli were their physical properties (e.g., pitch, loudness) and spatial locations. The test subjects were able to distinguish between the stimuli in approximately 70 % of the cases. If in-ear headphones were used instead of real speakers, the identification of target direction was very difficult. Rutkowski et al. (2010) extended the spatial sound design by also adding actual speakers to play the stimuli. They used seven speakers which could produce seven different BCI commands. The users were positioned in the middle of the 7.1 speaker system and were asked to concentrate on the direction of a single speaker and ignore the others. The target and non-target direction sequences were presented randomly. Their results confirmed that it is possible to determine specific components in the EEG signals evoked when concentrating on a selected sound.

The results of these studies are encouraging for further development of auditory and spatial auditory BCIs; however, their feasibility for real-world applications is currently rather limited due to limited accuracy and dependability. In addition, various studies have shown also that psychological factors such as motivation and mood also affect the BCI usability and performance in real-life applications.

4.8 Conclusion

The examples of spatial auditory interfaces reviewed in this chapter represent a special group of auditory interfaces that use the position of sounds and sound sources as additional informational cues in an interface. They offer an excellent alternative for exchanging information between a device and a user engaged in some sort of primary activity, such as walking, running, or driving, a user engaged in a virtual world or a user with visual impairment. Most of the prototypes of the advanced spatial auditory interfaces reviewed in this chapter are still not integrated into commercially-available products, but have proven to be very effective and useful in various situations and environments. The explained interaction techniques and the corresponding input and output modalities are expected to have a much stronger impact on future devices and environments, in which auditory interfaces will play an import role.

References

Air F (2004) Spatial audio display concepts supporting situation awareness for operators of unmanned aerial vehicles. In: Performing organization nameisi and addresiesi: 61

Augmented Reality (2014) Augmented Reality from Wikipedia. http://en.wikipedia.org/wiki/Augmented_reality. Accessed 31 Dec 2014

Azuma RT (1997) A survey of augmented reality. Presence 6(4):355–385

Beattie D, Baillie L, Halvey M, McCall R (2014) What's around the corner? Enhancing driver awareness in autonomous vehicles via in-vehicle spatial auditory displays. In: NordiCHI'14, ACM, p 10

Begault DR (1993) Head-up auditory displays for traffic collision avoidance system advisories: a preliminary investigation. Hum Factors J Hum Factors Ergon Soc 35(4):707–717

Begault DR, Pittman MT (1996) Three-dimensional audio versus head-down traffic alert and collision avoidance system displays. Int J Aviat Psychol 6(1):79–93

Bellotti F, Berta R, De Gloria A, Margarone M (2002) Using 3d sound to improve the effectiveness of the advanced driver assistance systems. Pers Ubiquit Comput 6(3):155–163

Billinghurst M, Bowskill J, Jessop M, Morphett J (1998) A wearable spatial conferencing space. In: Wearable computers, 1998. Digest of papers. Second international symposium, pp 76–83

Billinghurst M, Deo S, Adams N, Lehikoinen J (2007). Motion-tracking in spatial mobile audio-conferencing. In: Workshop on spatial audio for mobile devices (SAMD 2007) at Mobile HCI

Blommer M, Greenberg J (2003) Realistic 3D sound simulation in the VIRTTEX driving simulator. In: Proceedings of DSC North America

Brewster S, Lumsden J, Bell M, Hall M, Tasker S (2003) Multimodal 'eyes-free' interaction techniques for wearable devices. In: Proceedings of the SIGCHI conference on Human factors in computing systems, ACM, pp 473–480

Bronkhorst AW, Veltman JH, Van Breda L (1996) Application of a three-dimensional auditory display in a flight task. Hum Factors J Hum Factors Ergon Soc 38(1):23–33

Chang M, Nishikawa N, Struzik ZR, Mori K, Makino S, Mandic D, Rutkowski TM (2013) Comparison of P300 responses in auditory, visual and audiovisual spatial speller BCI paradigms. In: arXiv preprint arXiv:1301.6360

Chapman RM, Bragdon HR (1964) Evoked responses to numerical and non-numerical visual stimuli while problem solving. Nature 203:1155–1157

Cohen M, Aoki S, Koizumi N (1993) Augmented audio reality: Telepresence/VR hybrid acoustic environments. In: Proceedings of 2nd IEEE international workshop on robot and human communication, IEEE, pp 361–364

Crispien K, Fellbaum K, Savidis A, Stephanidis C (1996) A 3D-auditory environment for hierarchical navigation in non-visual interaction. In: Proceedings of ICAD, vol. 96, pp 4–6

Crispien K, Würz W, Weber G (1994) Using spatial audio for the enhanced presentation of synthesised speech within screen-readers for blind computer users. Springer, Berlin, pp 144–153

Deo S, Billinghurst M, Adams N, Lehikoinen J (2007) Experiments in spatial mobile audio-conferencing. In: Proceedings of the 4th international conference on mobile technology, applications, and systems and the 1st international symposium on Computer human interaction in mobile technology, ACM, pp 447–451

Dobler D, Haller M, Stampfl P (2002) ASR – Augmented Sound Reality. In: ACM SIGGRAPH 2002 conference abstracts and applications, p 148

Ekiga (2014) http://ekiga.org/. Accessed 31 Dec 2014

Farwell LA, Donchin E (1988) Talking off the top of your head: Toward a mental prosthesis utilizing event-related brain potentials. Electroencephalogr Clin Neurophysiol 70:512–523

Frauenberger C, Höldrich R, De Campo A (2004a) A generic, semantically-based design approach for spatial auditory computer displays. In: Proceedings of ICAD 04

Frauenberger C, Noistering M (2003) 3D audio interfaces for the blind

Frauenberger C, Putz V, Holdrich R (2004b) Spatial auditory displays—a study on the use of virtual audio environments as interfaces for users with visual disabilities. In: DAFx04 Proceedings, pp 5–8

Frauenberger C, Putz V, Holdrich R, Stockman T (2005) Interaction patterns for auditory user interfaces. In: ICAD proceedings Limerick Ireland, pp 154–160.

Frauenberger C, Stockman T (2006) Patterns in auditory menu design. In: Proceedings of the 12th international conference on auditory display (ICAD06), London, pp 141–147

Furdea A, Halder S, Krusienski DJ, Bross D, Nijboer F, Birbaumer N, Kübler A (2009) An auditory oddball (P300) spelling system for brain-computer interfaces. Psychophysiology 46(3): 617–625

Gerhard E (2001) Immersive audio-augmented environments: the LISTEN project. In: Proceedings of fifth international conference on information visualisation, IEEE, pp 571–573

Goose S, Kodlahalli S, Pechter W, Hjelsvold R (2002) Streaming speech 3: a framework for generating and streaming 3D text-to-speech and audio presentations to wireless PDAs as specified using extensions to SMIL. In: Proceedings of the 11th international conference on World Wide Web, ACM, pp 37–44

Grabianowski E (2007) How brain-computer interfaces work. HowStuffWorks. http://computer.howstuffworks.com/brain-computer-interface.htm. Accessed 31 Dec 2014

Graham R (1999) Use of auditory icons as emergency warnings: evaluation within a vehicle collision avoidance application. Ergonomics 42(9):1233–1248

Goose S, Möller C (1999) A 3D audio only interactive Web browser: using spatialization to convey hypermedia document structure. In: Proceedings of the seventh ACM international conference on Multimedia (Part 1), pp 363–371

Haller M, Dobler D, Stampfl P (2002) Augmenting the reality with 3D sound sources. In: ACM SIGGRAPH 2002 conference abstracts and applications, p 65

Haas EC (1998) Can 3-D auditory warnings enhance helicopter cockpit safety? In: Proceedings of the human factors and ergonomics society 42nd annual meeting, Santa Monica, CA: The Human Factors and Ergonomics Society, pp 1117–1121

Härmä A, Jakka J, Tikander M, Karjalainen M, Lokki T, Hiipakka J, Lorho G (2004) Augmented reality audio for mobile and wearable appliances. J Audio Eng Soc 52(6):618–639

Härmä A, Jakka J, Tikander M, Karjalainen M, Lokki T, Nironen H (2003) Techniques and applications of wearable augmented reality audio. In: Audio Engineering Society Convention 114, Audio Engineering Society

HMD (2014) Head-mounted display from Wikipedia. http://en.wikipedia.org/wiki/Head-mounted_display. Accessed 31 Dec 2014

Hiipakka J, Gaëtan L (2003) A spatial audio user interface for generating music playlists. In: Proceedings of the 2003 international conference on auditory display, pp 267–270

Hill NJ, Lal TN, Bierig K, Birbaumer N, Schölkopf B (2004) An auditory paradigm for brain-computer interfaces. In: Advances in neural information processing systems, pp 569–576

Ho C, Spence C (2005) Assessing the effectiveness of various auditory cues in capturing a driver's visual attention. J Exp Psychol Appl 11(3):157

Ho C, Spence C (2009) Using peripersonal warning signals to orient a driver's gaze. Hum Factors J Hum Factors Ergon Soc 51(4):539–556

Holland S, Morse DR, Gedenryd H (2002) AudioGPS: spatial audio navigation with a minimal attention interface. Pers Ubiquit Comput 6(4):253–259

Houtsma AJ (2003) Nonspeech audio in helicopter aviation (No. USAARL-2004-03) ARMY AEROMEDICAL RESEARCH LAB FORT RUCKER AL

Hyder M, Haun M, Hoene C (2009) Measurements of sound localization performance and speech quality in the context of 3D audio conference calls. In: Internation conference on acoustics, Rotterdam, Netherlands: NAG/DAGA

Hyder M, Haun M, Christian H (2010) Placing the participants of a spatial audio conference call. In: Consumer communications and networking conference (CCNC), pp 1–7

Jagdish D, Gupta M (2008) Sonic grid: an auditory interface for the visually impaired to navigate GUI-based environments. In: Proceedings of the IUI'08, Maspalomas, Gran Canaria, Spain, pp 337–340

Jakus G, Dicke C, Sodnik J (2015) A user study of auditory, head-up and multi-modal displays in vehicles. Appl Ergon 46:184–192

Jeon M, Davison BK, Nees MA, Wilson J, Walker BN (2009) Enhanced auditory menu cues improve dual task performance and are preferred with in-vehicle technologies. In: Proceedings of the 1st international conference on automotive user interfaces and interactive vehicular applications, pp 91–98

Kajastila R, Siltanen S, Lunden P, Lokki T, Savioja L (2007) A distributed real-time virtual acoustic rendering system for dynamic geometries. In 122nd convention of the audio engineering society (AES), Vienna, Austria

Kan A, Pope G, Jin C, van Shaik A (2004) Mobile spatial audio communication system. In: Proceedings of ICAD 04

Kanoh SI, Miyamoto KI, Yoshinobu T (2008) A brain-computer interface (BCI) system based on auditory stream segregation. In: EMBS 2008, 30th annual international conference of the IEEE, pp 642–645

Kilgore R, Chignell M, Smith P (2003) Spatialized audioconferencing: what are the benefits? In: Proceedings of the 2003 conference of the centre for advanced studies on collaborative research, IBM Press, pp 135–144

Kleiner M, Dalenbäck BI, Svensson P (1993) Auralization—an overview. J Audio Eng Soc 41(11):861–875

Kobayashi M, Schmandt C (1997) Dynamic Soundscape: mapping time to space for audio browsing. In: Proceedings of the ACM SIGCHI conference on human factors in computing systems, ACM

Krueger MW (1991) Artificial reality II (Vol. 10). Reading (MA), Addison-Wesley

Kyriakakis C (1998) Fundamental and technological limitations of immersive audio systems. Proc IEEE 86(5):941–951

Langlotz T, Regenbrecht H, Zollmann S, Schmalstieg D (2013) Audio stickies: visually-guided spatial audio annotations on a mobile augmented reality platform. In: Proceedings of the 25th Australian computer-human interaction conference: augmentation, application, innovation, collaboration, pp 545–554

Lentz T, Schröder D, Vorländer M, Assenmacher I (2007) Virtual reality system with integrated sound field simulation and reproduction. EURASIP J Appl Signal Proc 2007(1):187–187

Mariette N (2006) A novel sound localization experiment for mobile audio augmented reality applications, Advances in artificial reality and tele-existence. Springer, Berlin, pp 132–142

Martin A, Jin C, Schaik AV (2009) Psychoacoustic evaluation of systems for delivering spatialized augmented-reality audio. J Audio Eng Soc 57(12):1016–1027

Meijer PB (1992) An experimental system for auditory image representations. IEEE Trans Biomed Eng 39(2):112–121

Mereu SW, Kazman R (1996) Audio enhanced 3D interfaces for visually impaired users. In: Proceedings of the SIGCHI conference on human factors in computing systems, pp 72–78

Mourant RR, Refsland D (2003) Developing a 3d sound environment for a driving simulator. In: Proceedings of the ninth international conference on virtual systems and multimedia, pp 711–719

Mumble (2014) http://www.mumble.com/. Accessed 31 Dec 2014

Mynatt ED, Back M, Want R, Frederick R (1997) Audio Aura: light-weight audio augmented reality. In: Proceedings of the 10th annual ACM symposium on User interface software and technology, ACM, pp 211–212

Mynatt ED, Back M, Want R, Baer M, Ellis JB (1998) Designing audio aura. In: Proceedings of the SIGCHI conference on Human factors in computing systems, ACM, pp 566–573

Nakaizumi C, Matsui T, Mori K, Makino S, Rutkowski TM (2014) Head-related impulse response-based spatial auditory brain-computer interface. In: arXiv preprint arXiv:1404.3958

Nasir T, Roberts J (2007) Sonification of spatial data. In: Proceedings of the 13th international conference on auditory display (ICAD 2007), pp 112–119

Nijboer F, Furdea A, Gunst I, Mellinger J, McFarland DJ, Birbaumer N, Kübler A (2008) An auditory brain–computer interface (BCI). J Neurosci Methods 167(1):43–50

Oving AB, Veltman JA, Bronkhorst AW (2004) Effectiveness of 3-D audio for warnings in the cockpit. Int J Aviat Psychol 14(3):257–276

Parker SP, Smith SE, Stephan KL, Martin RL, McAnally KI (2004) Effects of supplementing head-down displays with 3-D audio during visual target acquisition. Int J Aviat Psychol 14(3):277–295

Pirhonen A, Brewster S, Holguin C (2002) Gestural and audio metaphors as a means of control for mobile devices. In: Proceedings of the SIGCHI conference on Human factors in computing systems, ACM, pp 291–298

Refreshable Braille Display (2014) Refreshable Braille Display from Wikipedia. http://en.wikipedia.org/wiki/Refreshable_braille_display. Accessed 31 Dec 2014

Regenbrecht H, Lum T, Kohler P, Ott C, Wagner M, Wilke W, Mueller E (2004) Using augmented virtuality for remote collaboration. Presence 13(3):338–354

Reynolds CJ, Reed MJ, Hughes PJ (2008) Analysis of a distributed processing model for spatialized audio conferences. In: 2008 IEEE international conference on multimedia and expo, pp 461–464

Reynolds CJ, Reed MJ, Hughes PJ (2009) Decentralized headphone based spatialized audio conferencing for low power devices. In: International conference on multimedia and expo, ICME 2009, IEEE, pp 778–781

Roth P, Petrucci LS, Assimacopoulos A, Pun T (2000) Audio-haptic internet browser and associated tools for blind and visually impaired computer users. In: Workshop on friendly exchanging through the net, pp 22–24

Roth P, Petrucci L, Pun T, Assimacopoulos A (1999) Auditory browser for blind and visually impaired users. In: CHI'99 extended abstracts on Human factors in computing systems, pp 218–219

Rothbucher M, Habigt T, Feldmaier J, Diepold K (2010) Integrating a HRTF-based sound synthesis system into Mumble. In: 2010 IEEE international workshop on multimedia signal processing (MMSP), pp 24–28

Rothbucher M, Kaufmann M, Habigt T, Feldmaier J, Diepold K (2011). Backwards compatible 3D audio conference server using HRTF synthesis and SIP. In: Seventh international conference on signal-image technology and internet-based systems (SITIS), pp 111–117

Rutkowski TM, Tanaka T, Zhao Q, Cichocki A (2010) Spatial auditory BCI/BMI paradigm-Multichannel EMD approach to brain responses estimation. In: APSIPA annual summit and conference, pp 197–202

Sánchez J, Sáenz M (2006) 3D sound interactive environments for blind children problem solving skills. Behav Inform Technol 25(4):367–378

Sánchez J, Sáenz M (2010) Metro navigation for the blind. Comput Educ 55(3):970–981

Savioja L, Huopaniemi J, Lokki T, Väänänen R (1999) Creating interactive virtual acoustic environments. J Audio Eng Soc 47(9):675–705

Sawhney N, Schmandt C (2000) Nomadic radio: speech and audio interaction for contextual messaging in nomadic environments. ACM Trans Comput Hum Interact (TOCHI) 7(3):353–383

Schmandt C (1995) AudioStreamer: exploiting simultaneity for listening. In: Proceedings of the computer-human interaction (CHI), pp 218–219

Schmandt C (1998) Audiohallway: a virtual acoustic environment for browsing. In: Proceedings of the 11th annual ACM symposium on user interface software and technology (UIST'98), San Francisco,CA, pp 163–170

Schreuder M, Blankertz B, Tangermann M (2010) A new auditory multi-class brain-computer interface paradigm: spatial hearing as an informative cue. PLoS One 5(4), e9813

Screen reader (2014) Screen reader from Wikipedia. http://en.wikipedia.org/wiki/Screen_reader. Accessed 31 Dec 2014

Shengdong Z, Dragicevic P, Chignell M, Balakrishnan R, Baudisch P (2007) Earpod: eyes-free menu selection using touch input and reactive audio feedback. In: Proceedings of the SIGCHI conference on Human factors in computing systems, ACM, pp 1395–1404

Smil (2014) Synchronized Multimedia Integration Language from Wikipedia. http://en.wikipedia.org/wiki/Synchronized_Multimedia_Integration_Language. Accessed 31 Dec 2014

Sodnik J, Dicke C, Tomažič S, Billinghurst M (2008) A user study of auditory versus visual interfaces for use while driving. Int J Hum Comput Stud 66(5):318–332

Sodnik J, Jakus G, Tomažič S (2011) Multiple spatial sounds in hierarchical menu navigation for visually impaired computer users. Int J Hum Comput Stud 69(1):100–112

Sodnik J, Jakus G, Tomažič S (2012) The use of spatialized speech in auditory interfaces for computer users who are visually impaired. J Vis Impair Blindness 106(10):634–645

Sodnik J, Tomažič S (2011) Spatial speaker spatial positioning of synthesized speech in Java. Lect Notes Electr Eng 68:359–371

Sodnik J, Tomazic S, Grasset R, Duenser A, Billinghurst M (2006) Spatial sound localization in an augmented reality environment. In: Proceedings of the 18th Australia conference on computer-human interaction: design: activities, artefacts and environments, pp 111–118

Stampfl P (2003a) Augmented reality disk jockey: AR/DJ. In: Proceedings of SIGGRAPH 2003, p 1

Stampfl P (2003b) 3deSoundBox – a scalable, platform-independent 3D sound system for virtual and augmented reality applications. In: Proceedings of EUROGRAPHICS 2003

Sundareswaran V, Wang K, Chen S, Behringer R, McGee J, Tam C, Zahorik P (2003) 3D audio augmented reality: implementation and experiments. In: Proceedings of the 2nd IEEE/ACM international symposium on mixed and augmented reality, p 296

Susnik R, Sodnik J, Tomazic S (2008) An elevation coding method for auditory displays. Appl Acoust 69(3):233–241

Suzuki K, Jansson H (2003) An analysis of driver's steering behaviour during auditory or haptic warnings for the designing of lane departure warning system. JSAE Rev 24(1):65–70

TCAS (2014) Traffic collision avoidance system from Wikipedia. http://en.wikipedia.org/wiki/Traffic_collision_avoidance_system. Accessed 31 Dec 2014

Teleconference (2014) Teleconference from Wikipedia. http://en.wikipedia.org/wiki/Teleconference. Accessed 31 Dec 2014

Tikander M, Härmä A, Karjalainen M (2003) Binaural positioning system for wearable augmented reality audio. In: Workshop on applications of signal processing to audio and acoustics, IEEE, pp 153–156

Tonnis M, Klinker G (2006) Effective control of a car driver's attention for visual and acoustic guidance towards the direction of imminent dangers. In: Proceedings of the 5th IEEE and ACM international symposium on mixed and augmented reality, pp 13–22

Vazquez-Alvarez Y, Oakley I, Brewster SA (2012) Auditory display design for exploration in mobile audio-augmented reality. Pers Ubiquit Comput 16(8):987–999

Vazquez-Alvarez Y, Aylett MP, Brewster SA, von Jungenfeld R, Virolainen A (2014) Multilevel auditory displays for mobile eyes-free location-based interaction. In: CHI'14 extended abstracts on human factors in computing systems, ACM, pp 1567–1572

Veltman JA, Oving AB, Bronkhorst AW (2004) 3-D Audio in tlie fighter coclcpit improves task performance

Virtual Reality (2014) Virtual Reality from Wikipedia. http://en.wikipedia.org/wiki/Virtual_reality. Accessed 31 Dec 2014

VoIP (2014) Voice over IP from Wikipedia. http://en.wikipedia.org/wiki/Voice_over_IP. Accessed 31 Dec 2014

Walker A, Brewster S (2000) Spatial audio in small screen device displays. Pers Technol 4(2–3):144–154

Walker A, Brewster S, McGooking D, Ng A (2001) Diary in the sky: a spatial audio display for a mobile calendar. People and Computers XV—Interaction without Frontiers, Springer, London, pp 531–539

Wakkary R, Hatala M (2007) Situated play in a tangible interface and adaptive audio museum guide. Pers Ubiquit Comput 11(3):171–191

Wolpaw JR, Birbaumer N, McFarland DJ, Pfurtscheller G, Vaughan TM (2002) Brain–computer interfaces for communication and control. Clin Neurophysiol 113(6):767–791

Zahorik P, Tam C, Wang K, Bangayan P, Sundareswaran V (2001) Localization accuracy in 3-D sound displays: the role of visual-feedback training. In: Proceedings of the advanced displays and interactive displays federal laboratory consortium

Zhou Z, Cheok AD, Yang X, Qiu Y (2004) An experimental study on the role of 3D sound in augmented reality environment. Interact Comput 16(6):1043–1068

Printed in the United States
By Bookmasters